# The Future of Post-Human Urban Planning

# The Future of Post-Human Urban Planning

## A Preface to a New Theory of Density, Void, and Sustainability

By

Peter Baofu

CAMBRIDGE SCHOLARS PUBLISHING

The Future of Post-Human Urban Planning: A Preface to a New Theory of Density, Void, and Sustainability, by Peter Baofu

This book first published 2009

Cambridge Scholars Publishing

12 Back Chapman Street, Newcastle upon Tyne, NE6 2XX, UK

British Library Cataloguing in Publication Data
A catalogue record for this book is available from the British Library



ISBN (10): 1-4438-0591-2, ISBN (13): 978-1-4438-0591-9

To Those in the World Beyond Sustainability

# BOOKS ALSO BY PETER BAOFU

**Category I: Social Sciences and Related Fields**

- 22. *The Future of Post-Human Mass Media* (2009)
- 21. *The Rise of Authoritarian Liberal Democracy* (2007)
- 20. *Beyond the World of Titans, and the Remaking of World Order* (2007)
- 19. *Beyond Capitalism to Post-Capitalism* (2005)
- 18. Volume 1: *Beyond Democracy to Post-Democracy* (2004)
- 17. Volume 2: *Beyond Democracy to Post-Democracy* (2004)
- 16. *The Future of Capitalism and Democracy* (2002)

**Category II: Natural Sciences and Related Fields**

- 15. *The Future of Post-Human Engineering* (2009)
- 14. *The Future of Post-Human Unconsciousness* (2008)
- 13. *The Future of Complexity* (2007)
- 12. *Beyond Nature and Nurture* (2006)
- 11. *The Future of Post-Human Space-Time* (2006)
- 10. *The Future of Post-Human Consciousness* (2004)

**Category III: Formal Sciences and Related Fields**

- 9. *The Future of Post-Human Geometry* (2009)
- 8. *The Future of Post-Human Mathematical Logic* (2008)
- 7. *The Future of Information Architecture* (2008)

**Category IV: The Humanities and Related Fields**

- 6. *The Future of Post-Human Creative Thinking* (2009)
- 5. *The Future of Post-Human Knowledge* (2008)
- 4. *The Future of Aesthetic Experience* (2007)
- 3. *Beyond Civilization to Post-Civilization* (2006)
- 2. Volume 1: *The Future of Human Civilization* (2000)
- 1. Volume 2: *The Future of Human Civilization* (2000)

# CONTENTS

## Part Three: Urban Void



## Part Four: Conclusion

# TABLES

*Category III: Visions on the Mind*



*Category IV: Visions on History*

# FOREWORD

For readers who are encountering Dr. Peter Baofu for the first time with this volume, it may come as a surprise to learn about the range of subjects he has addressed in preceding works. He has an insatiable curiosity for questioning theoretical frameworks and takes a disciplined approach to examining what other researchers have said on a subject.

In the present volume he questions the tenets of existing thinking about urban planning and posits a new approach to the subject. His ideas are insightful, fresh and value laden. They present a new challenge for city planners, legislators and community action groups. They present a blueprint for better living in a city of tomorrow.

Dr. Baofu prods the reader to move forward with an urban planning process and offers some helpful guideposts to lead the way.

*Sylvan Von Burg*
*School of Business*
*George Washington University*

# ACKNOWLEDGMENTS

This book, like all previous ones of mine, is written to challenge contemporary wisdom, this time, on density, void, and sustainability—and then to propose a new theory to understand the subject matter better.

Because of this political incorrectness, this book receives no external funding nor help from any formal organization or institution.

The only reward that I receive is the wonderful feeling to discover something new to say in a way not proposed before.

There is one person, however, whom I deeply appreciate for his foreword, and he is Sylvan von Burg at George Washington University School of Business,.

In any event, I bear the sole responsibility for the ideas presented in this book.

# ABBREVIATIONS

ALD = Peter Baofu. 2007. *The Rise of Authoritarian Liberal Democracy: A Preface to a New Theory of Comparative Political Systems*. Cambridge, England: Cambridge Scholars Publishing, Ltd.

BCIV = Peter Baofu. 2006. *Beyond Civilization to Post-Civilization: Conceiving a Better Model of Life Settlement to Supersede Civilization*. NY: Peter Lang Publishing, Inc.

BCPC = Peter Baofu. 2005. *Beyond Capitalism to Post-Capitalism: Conceiving a Better Model of Wealth Acquisition to Supersede Capitalism*. NY: The Edwin Mellen Press.

BDPD1 = Peter Baofu. 2004. Volume 1. *Beyond Democracy to Post-Democracy: Conceiving a Better Model of Governance to Supersede Democracy*. NY: The Edwin Mellen Press.

BDPD2 = Peter Baofu. 2004. Volume 2. *Beyond Democracy to Post-Democracy: Conceiving a Better Model of Governance to Supersede Democracy*. NY: The Edwin Mellen Press.

BNN = Peter Baofu. 2006. *Beyond Nature and Nurture: Conceivng a Better Way to Understand Genes and Memes.* Cambridge, England: Cambridge Scholars Publishing, Ltd.

BWT = Peter Baofu. 2007. *Beyond the World of Titans, and the Renaking of World Order: A Preface to a New Logic of Empire-Building.* Cambridge, England: Cambridge Scholars Publishing, Ltd.

FAE = Peter Baofu. 2007. *The Future of Aesthetic Experience: Conceiving a Better Way to Understand Beauty, Ugliness and the Rest.* Cambridge, England: Cambridge Scholars Publishing, Ltd.

FC = Peter Baofu. 2007. *The Future of Complexity: Conceiving a Better Way to Understand Order and Chaos.* London, United Kingdom: World Scientific Publishing Co.

FCD = Peter Baofu. 2002. *The Future of Capitalism and Democracy.* MD: The University Press of America.

FHC1 = Peter Baofu. 2000. Volume 1. *The Future of Human Civilization.* NY: The Edwin Mellen Press.

FHC2 = Peter Baofu. 2000. Volume 2. *The Future of Human Civilization.* NY: The Edwin Mellen Press.

FIA = Peter Baofu. 2008. *The Future of Information Architecture: Conceiving a Better Way to Understand Taxonomy, Network, and Intelligence.* Oxford, England: Chandos Publishing (Oxford) Limited.

FPHC = Peter Baofu. 2004. *The Future of Post-Human Consciousness.* NY: The Edwin Mellen Press.

FPHCT = Peter Baofu. 2009. *The Future of Post-Human Creative Thinking: A Preface to a New Theory of Invention and Innovation.* Cambridge, England: Cambridge Scholars Publishing, Ltd.

FPHE = Peter Baofu. 2009. *The Future of Post-Human Engineering: A Preface to a New Theory of Technology.* Cambridge, England: Cambridge Scholars Publishing, Ltd.

FPHG = Peter Baofu. 2009. *The Future of Post-Human Geometry: A Preface to a New Theory of Infinity, Symmetry, and Dimensionality.* Cambridge, England: Cambridge Scholars Publishing, Ltd.

FPHK = Peter Baofu. 2008. *The Future of Post-Human Knowledge: A Preface to a New Theory of Methodology and Ontology.* Oxford, England: Chandos Publishing (Oxford) Limited.

FPHML = Peter Baofu. 2008. *The Future of Post-Human Mathematical Logic: A Preface to a New Theory of Rationality.* Cambridge, England: Cambridge Scholars Publishing, Ltd.

FPHMM = Peter Baofu. 2009. *The Future of Post-Human Mass Media: A Preface to a New Theory of Technology.* Cambridge, England: Cambridge Scholars Publishing, Ltd.

FPHST = Peter Baofu. 2006. *The Future of Post-Human Space-Time: Conceivng a Better Way to Understand Space and Time.* New York: Peter Lang Publishing, Inc.

FPHU = Peter Baofu. 2008. *The Future of Post-Human Unconsciousness: A Preface to a New Theory of Anomalous Experience.* Cambridge, England: Cambridge Scholars Publishing, Ltd.

FPHUP = Peter Baofu. 2009. *The Future of Post-Human Urban Planning: A Preface to a New Theory of Density, Void, and Sustainability.* Cambridge, England: Cambridge Scholars Publishing, Ltd.

# • Part One •

## *Introduction*

# CHAPTER 1
# *INTRODUCTION—THE LEGACY OF URBAN PLANNING*

> [W]e have gradually come to see planning as a much broader set of human activities, encompassing the physical world and also the realm of public and social services. While retaining technical analytical and design components, planning has come to be seen also as intensely political and value laden.
>
> —M. Wachs (1995)

## The Obsession with Sustainability

Why should urban planning in our time be obsessed with the issue of sustainability? Or differently put, is sustainability really as desirable and possible as its proponents in urban planning (and other related fields like economics, political science, environmental studies, architecture, and so on) would like us to believe?

Contrary to the conventional wisdom held by many since the modern era, the concern with sustainability has been much exaggerated and distorted, to the point that it is fast becoming a new intellectual fad, so that its dark sides have been unwarrantedly ignored or downgraded.

This is not to say, however, that the literature on sustainability in urban planning (and other related fields) hitherto existing in history has been full of nonsense. Indeed, on the contrary, much can be learned from different theoretical approaches in the literature.

The important point to remember here, however, is that this book provides an alternative (better) way to understand the nature of sustainability in urban planning (and other related fields), which learns from different sides of the debate but in the end transcends them all.

The urgency of this inquiry should not be underestimated, as it concerns not only urban planning (as a case study here) but also other highly related yet very serious challenges in our time (e.g., ecological, economic, demographic, technological, moral, spiritual, political, and the like).

Therefore, if true, this seminal view will fundamentally change the way that we think about the issue of sustainability, with its enormous implications not only for understanding the future of urban planning, in a small sense—but also for predicting the relevance of sustainability in relation to the entire domain of human knowledge for the human future and what I originally called its "post-human" fate, in a broad sense.

## A Metamorphic History of Urban Planning

A good way to start is to define the words "urban planning," with the issue of sustainability as the main focus.

Firstly, the term "urban" refers to a city, which, as A.E. J. Morris (1972) once put it, is "a community of substantial size and population density that shelters a variety of nonagricultural specialists, including a literate elite…by adding a number of other requirements: production of a surplus of food by one section of the society; existence of some form of writing; social organisation to ensure continuity of supplies to the urban specialists; and technological expertise to transport and store materials and to undertake civic works."

Of course, this definition does not mean that the "transition from village/town to city" is "a clearly definable process. Even a primitive tribal compound has many of the elements of a city in embryo form." (D. Engwicht 2009)

And secondly, the term "planning" is often contrasted with something which is "organic," with its "chaotic street pattern that has resulted from ad hoc placement of buildings, streets being defined by the placement of buildings over time." (D. Engwicht 2009)

By contrast, a "planned" (not "organic") city has "a recognizable pattern to the street network, usually a gridiron. In these cities, the streets define where the buildings are placed, and not vice versa. (D. Engwicht 2009; A. Morris 1972; C. Cunningham 1992)

Of course, this distinction should not be taken too rigidly, since "it is possible that the so called organic cities were also planned to some degree or another." (D. Engwicht 2009)

This definition of the words "urban planning" aside—the next business to do is to introduce a brief history of urban planning, which is metamorphic enough through the ages, that is, in different stages of development from antiquity to our time.

Good examples in this brief history include, say, (a) the "grid plan," (b) the "fortress" model, (c) the "star-shaped city," (d) the "garden city," (e) the "modernist city," and (f) the "post-modernist" model—as briefly summarized below.

## The "Grid Plan"

In antiquity, a well-known example of urban planning in its earliest stage involves what is now known as "the grid plan or gridiron plan," which "is a type of city plan in which streets run at right angles to each other, forming a grid. In the context of the culture of Ancient Greece the grid plan is called Hippodamian plan." (WK 2008c; K. Jackson 1985)

The "Hippodamian plan" (for the "grid plan") is named after the Greek urban planner known as Hippodamus of Miletus (498 B.C.—408 B.C.), who is often regarded as the "father" of urban planning in the Western world. (WK 2008e)

The grid plan dated from antiquity also existed in different cultures. For instance, in India, by "2600 B.C., Mohenjo-daro and Harappa, major cities of the Indus Valley Civilization, were built with blocks divided by a grid of straight streets, laid out in perfect right angles, running north-south and east-west. Each block was subdivided by small lanes." (WK 2008c)

In Babylon, by the $17^{th}$ century B.C., when "Hammurabi…was a king of the Babylonian Empire who made Babylon one of the greatest metropolises in antiquity. He rebuilt Babylon, building and restoring temples, city walls, public buildings, and building canals for irrigation. The streets of Babylon were wide and straight, intersected approximately at right angles, and were paved with bricks and bitumen." (WK 2008c)

In ancient China, "from the 15th century B.C. onward…[g]uidelines put into written form in the *Kaogong ji* during the Spring and Autumn Period (770-476 BC) stated: 'a capital city should be square on plan. Three gates on each side of the perimeter lead into the nine main streets that crisscross the city and define its grid-pattern. And for its layout the city should have the Royal Court situated in the south, the Marketplace in the north, the Imperial Ancestral Temple in the east and the Altar to the Gods of Land and Grain in the west.'" (WK 2008c)

And in ancient Rome, the grid plan was further developed, this time, with the focus on "military defense and civil convenience," in that the

Roman grid plan "is a central forum with city services, surrounded by a compact rectilinear grid of streets and wrapped in a wall for defense. To reduce travel times, two diagonal streets cross the square grid corner-to-corner, passing through the central square. A river usually flowed through the city, to provide water, transport, and sewage disposal." (WK 2008b)

## The "Fortress" Model

But in Medieval Europe, urban planning was shifted to a different focus, this time "on a fortress, a fortified abbey, or a (sometimes abandoned) Roman nucleus,…'like the annular rings of a tree,' whether in an extended village or the center of a larger city. Since the new center was often on high, defensible ground, the city plan took on an organic character, following the irregularities of elevation contours like the shapes that result from agricultural terracing." (WK 2008b; S. Giedion 1941)

A distinctive feature of these medieval fortresses involves castle towers, which "were usually made of stone or sometimes (but rarely) wood. Often toward the later part of the era they included battlements and arrow loops. Arrow loops were vertical slits in the wall where archers from the inside shot arrows through at the attackers, but they made it extremely difficult for attackers to get many arrows through back at the defenders." (2008e)

But the medieval urban planning was much limited in further development, because of "the juridical chaos of medieval cities (where the administration of streets was sometimes hereditary with various noble families), and the characteristic tenacity of medieval Europeans in legal matters, prevented frequent or large-scale urban planning." (WK 2008b)

Another reason for the decline of the medieval "fortress" model has to do with the invention of cannons in the battlefields, which made the medieval fortresses vulnerable to cannon fires.

It was not until "the Renaissance and the enormous strengthening of all central governments, from city-states to the kings of France, characteristic of that epoch," that a new model of urban planning emerged. (WK 2008b)

## The "Star-Shaped City"

By the time of the Renaissance, there was the new model of urban planning for the "star-shaped city."

When contrasted with the "fortress" model—the "star-shaped city" was conceived to confront the threat of cannon shots, which came to dominate the battlefields at the time, since "[b]y that time, the medieval walls [in the "fortress" model] were no longer secure against a serious threat from an army as they were not designed for resisting cannon shot." (WK 2008e)

In other words, the "star-shaped city" was constructed "during the age of black powder, when cannons came to dominate the battlefield, and was first seen [during the] mid-15th century in Italy. Passive ring-shaped (enceinte) fortifications of the medieval era proved vulnerable to damage or destruction by cannon-fire, when it could be directed from outside against a perpendicular masonry wall." (WK 2008f)

By contrast, "the star fortress was a very flat structure composed of many triangular bastions, specifically designed to cover each other, and a ditch. Further structures such as ravelins, hornworks or crownworks, and detached forts could be added to create a complex symmetrical structure." (WK 2008f)

For instance, "Florence was an early model of the new urban planning, which rearranged itself into a star-shaped layout adapted from the new star fort, designed to resist cannon fire. This model was widely imitated, reflecting the enormous cultural power of Florence in this age; '[t]he Renaissance was hypnotized by one city type which for a century and a half—from Filarete to Scamozzi—was impressed upon utopian schemes: this is the star-shaped city.' Radial streets extend outward from a defined center of military, communal or spiritual power." (WK 2008b; S. Giedion 1941)

## The "Garden City"

Then came the "garden city" in the 19th century and later, when the Industrial Revolution reached its more advanced stage.

It is Sir Ebenezer Howard in the United Kingdom, who was inspired by Edward Bellamy's utopian novel titled *Looking Backward* (published in 1888) and, subsequently, proposed the idea of the "garden city" in *To-Morrow: A Peaceful Path to Real Reform* in 1898 (which was "reissued in 1902 as *Garden Cities of To-Morrow*"). (WK 2008g)

Each of these "garden cities" was "to be planned, self-contained communities surrounded by greenbelts, and containing carefully balanced areas of residences, industry, and agriculture." (WK 2008g) And each greenbelt "is an undeveloped area neighbouring an urban area, often protected from development by planning law." (WK 2008h)

This new model of urban planning is not surpirsing, since, by then, the "control of building was largely held by businesses and the wealthy elite.

Around 1900, there began to be a movement for providing citizens, especially factory workers, with healthier environments." (WK 2008h)

In fact, some of the well-known models of the "garden city" were later constructed, and good examples include "Letchworth and Welwyn Garden City, the world's first garden cities, in Hertfordshire, UK. However, these were principally small scale in size, typically dealing with only a few thousand residents." (WK 2008b; P. Hall 1998)

## The "Modernist City"

By the 1920's, there emerged the "modernist city"—especially though not exclusively for the building of "skyscrapers," partly because of the increase of population.

Modernism, in this sense, was "[b]ased on the ideas of Le Corbusier and utilising new skyscraper building techniques," to build "the modernist city [which] stood for the elimination of disorder, congestion and the small scale, replacing them instead with preplanned and widely spaced freeways and tower blocks set within gardens. There were plans for large scale rebuilding of cities, such as the Plan Voisin (based on Le Corbusier's Ville Contemporaine), which proposed clearing and rebuilding most of central Paris." (WK 2008b)

An exemplary model of the "modernist city" is undoubtedly Brasilia in Brazil, which "was planned and developed in 1956 with Lúcio Costa as the principal urban planner and Oscar Niemeyer as the principal architect. In 1960, it formally became Brazil's national capital. When seen from above, the main planned part of the city's shape resembles an airplane or a butterfly." (WK 2008i; GL 2008)

Brasilia became "a world reference for [modernist] urban planning. The locating of residential buildings around expansive urban areas, of building the city around large avenues and dividing it into sectors, has sparked a debate and reflection on life in big cities in the 20th century. The city's planned design included specific areas for almost everything, including accommodation—Hotel Sectors North and South." (WK 2008i)

But the ideal of the "modernist city" did not survive long, as it "was expressed in high crime and social problems within many of these planned neighbourhoods. Modernism can be said to have ended in the 1970s when the construction of the cheap, uniform tower blocks ended in many countries, such as Britain and France." (WK 2008b; E. Morris 1997)

### The "Postmodernist" Model

Therefore, by the 1970's, the model of "modernist city" gave way to its "post-modernist" successor, which stressed "individualism" and "diversity." (WK 2008b)

By then, "more conventional housing has been built. Rather than attempting to eliminate all disorder, planning now concentrates on individualism and diversity in society and the economy....Minimally-planned cities still exist." (WK 2008b)

A good illustration of "postmodernist" urban planning is Houston in Texas, since it "is an example of a large city (with a metropolitan population of 5.5 million) in a developed country, without a comprehensive zoning ordinance." (WK 2008b)

This is true, even though "Houston does, however, have many of the land use restrictions covered by traditional zoning regulations, such as restrictions on development density and parking requirements, even though specific land uses are not regulated. Moreover, private-sector developers in Houston have used subdivision covenants and deed restrictions effectively to create the same kinds of land use restrictions found in most municipal zoning laws." (WK 2008b)

But the "postmodernist" model does not require any major city to be like Houston, since it values more "diversity" and "individualism" in urban planning.

## Density, Void, and Sustainability

Surely, the introduction to the metamorphic history of urban planning (in the previous section) constitutes only a simplified brief history, with the sole purpose to serve as a starting point for the inquiry of the issue of sustainability in urban planning (and other related fields).

That qualified—the brief history also shows us something interesting, in that, although different models of urban planning come and go through the ages, they all share something in common, namely, in coping with the recurrent problems of some dualities.

Good examples include, say, (a) form (e.g., aesthetic appeal) and function (e.g., military defense), (b) order (e.g., noticeable patterns in street arrangement) and chaos (e.g., "the juridical chaos of medieval cities," and the decay in city formation over time), (c) simple networks (e.g., the plain grid plan for a small city) and complicated networks (e.g., complicated urban planning for a metapolis like Brasilia), (d) conventions (e.g., the conventional construction of an "Altar to the Gods" in a specified sector of an ancient Chinese city) and creative thinking (e.g, novel challenges to the status quo by subsequent alternative new models), (e) symmetry (e.g., the symmetrical structure in a "star-shaped city") and asymmetry (e.g., broken balance in a postmodernist world of diversity), and (f) density (e.g., highly populated areas) and void (e.g., urban decay and desertion).

(a) In relation to form and function, I already addressed the problems associated with this duality in *The Future of Aesthetic Experience* (*FAE*). And in *The Future of Post-Human Engineering* (*FPHE*), I also analyzed this problematic duality, especially from the vantage point of "technical" interests (e.g., the concern with efficiency) and "normative" interests (e.g., political consideration, aesthetic interest, and so on).

(b) In relation to order and chaos, I likewise already examined the problems of this duality in *The Future of Complexity* (*FC*).

(c) In relation to simple and complicated networks, I also addressed the problems associated with this duality in *The Future of Information Architecture* (*FIA*).

(d) In relation to conventions and creative thinking, I already looked into the problems related to this duality in *The Future of Post-Human Creative Thinking* (*FPHCT*).

(e) In relation to symmetry and asymmetry, I likewise already analyzed the problems associated with this duality in *The Future of Post-Human Geometry* (*FPHG*).

And (f), in relation to density and void, I shall address the problems of this duality in this book, that is, *The Future of Post-Human Urban Planning* (*FPHUP*).

In other words, I shall try to not repeat the analysis of the other related dualities (as already analyzed in my previous books), although all of them are relevant to the study of urban planning (and other related fields) too.

Instead, my focus here is on a different structural relationship, or what I want to call, in the absence of better words, (i) *density* and (ii) *void*, in the context of understanding the nature of (iii) *sustainability* in urban planning (and other related fields)—to be explicated in what follows.

## Density

The term "density," in its conventional usage, refers to "the quantity per unit volume, unit area, or unit length," "the distribution of a quantity…per unit usually of space (as length, area, or volume)," or "the average number of individuals or units per space unit." (MWD 2008)

In the context of urban planning, "density" is used here to refer to the distribution of entities (e.g., inhabitants, buildings, roads, parks, etc.) in a given city (or area, in general) over time.

The issue of "density," when contextualized in relation to sustainability in urban planning, concerns different phenomena (e.g., "traffic density, road rage, and human emotion"; "urban density and mental health"; "ecological complexity, and the urban balance with nature"; "urban pollution and institutional failures"; "urban renewal and structural injustice"; "sustainable technology and system complexity"; "urban planning and the religious search for order and purity"; "aesthetics and soundscape planning"; and the like).

Problems like this will be considered in Chapter Two.

## Void

By contrast, the term "void," in its conventional usage, refers to something which is "not occupied," "not inhabited" (or "deserted"), or simply "empty," relatively speaking, of course. (MWD 2008a)

This qualification, namely, "relatively speaking" above, is used with good reason, since, in physics, it is well known that "empty" space is not really empty all the way, because it can be full of energy (e.g., random quantum fluctuations at the sub-atomic level, and, for that matter, dark energy in the universe) and matter (e.g., different versions of sub-atomic particles, and, for that matter, dark matter in the universe). (F. Wilczek 2008)

Therefore, "it is practically impossible to construct a region of space that contains no matter or fields, since gravity cannot be blocked and all objects at a non-zero temperature radiate electromagnetically." (WK 2009j)

With this qualication in mind—the lack of density is "void," even if used here merely as an "approximation," depending on the degree of the lack of density.

In the context of urban planning, "void" here therefore means the relative absence of the distriubtion of entities (e.g., inhabitants, buildings, roads, parks, etc.) in a city (or area, in general) over time, such that it is not occupied, not inhabited, or empty—again, relatively speaking.

The issue of "void," when contextualized in relation to sustainability in urban planning, then refers to different phenomena (e.g., "indifference and urban solitude"; "urban homelessness and mental Health" ; "deforestation and urban destructiveness": "urban open space and its controversy"; "urban decay and its cycle"; "urbicide and urban warfare";

"individualism and urban emptiness"; "the monastic emptiness, and urban suffering"; and so on).

Problems like this will be analyzed in Chapter Three.

## Sustainability

The analysis of "density" and "void," therefore, can tell us something about the nature of sustainability in urban planning (and other related fields) over time.

The term "sustainability," in its conventional usage, refers to "a method of harvesting or using a resource so that the resource is not depleted or permanently damaged" over time. (WMD 2008b)

In the technical jargons of economics, for instance, sustainability strives for "optimization" (for the first best solution, or something close to it), not just "maximization" (for the second best solution, or something close to it, without dealing with, say, negative externalities like environmental pollution).

But Herbert Simon had famously criticised the talk of "optimality" and opted instead for "bounded rationality," in that "satisficing…is a decision-making strategy which attempts to meet criteria for adequacy, rather than to identify an optimal solution," and the reason is that "human beings lack the cognitive resources" to optimize, since "we usually do not know the relevant probabilities of outcomes, we can rarely evaluate all outcomes with sufficient precision, and our memories are weak and unreliable. A more realistic approach to rationality takes into account these limitations." (WK 2009n)

In contemporary (neo-Darwinian) evolutionary theory, many biologists now notice that "natural selection" is often suboptimal, because of "selective compromises," "phylogenetic inertia," or other causes. (P. Baofu 2006; D. Campbell 1975: 1120; M. Sahlins 1976: 78) In fact, many scientists also recognize that "evolution by natural selection often yields suboptimal but adequate design." (R. M. Alexander 2001:591) So, the classical Darwinian model of "survival of the fittest" (for "producing offsprings fast") is no longer as persuasive as it used to be; instead, there is now the contemporary (neo-Darwinian) version of "survival of the flattest" (for "producing offsprings fit"). (C. Wilke 2001; WP 2001:A9; P. Baofu 2002 & 2006).

These criticisms of "optimality" aside—in the current context of urban planning (and other related fields), "sustainability" has a technical meaning, in relation to four main features, namely, (a) temporal persistence, (b) ecological interdependence, (c) moral intergenerational equity, and (d) policy implications. (M. Dempster 1998)

(a) In relation to temporal persistence—the concern here is with "the persistence of the human species" (or its ability to exist over a long period of time). Different scholars disagree on the time frame in question, as it can mean, for instance, "indefinitely" for R. Costanza (1992), a "millennia into the future" for J. Pezzey (1992), the "correct social time horizon" for J. Dixon (1989) and L. Fallon, or no universal "temporal scale" at all, depending on "a system under consideration," for R. Costanza (1995) and B. Patten. (M. Dempster 1998)

(b) In relation to ecological interdependence—the issue here involves the linkage between human ability to exist over time and "the ecological processes necessary to meet such an objective," to the extent that there is a "recognition of human systems' connection to, and dependence upon, the natural environment." (M. Dempster 1998; J. Robinson 1990; R. Costanza 1992; J. Karr 1992)

(c) In relation to moral intergenerational equity—the issue here concerns "the moral principle of intergenerational equity," such that it is important for the present generation to act accordingly for the persistent existence of future posterity, in relation to the issues of "social values and quality of life," in conjunction with the support of nature. (M. Dempster 1998; S. Dovers (1992) and J. Handmer; R. Costanza 1992; J. Pezzey 1992; C. Holling 1995)

And (d), in relation to policy implications—it becomes obvious then that society, especially under the leadership of the state, should therefore play a major role to enforce coherent policies for a sustainable future of urban planning, so that, as M. Schwartz and M. Thompson strongly put it, "[w]hat we do today largely depends on how we interpret the past, and our interpretation of the past will, to a considerable extent, be shaped by the futures that our desires have already created." (M. Dempster 1998; L. Hempel 1996)

In fact, other scholars go a step further and argue that sustainability should also include other serious issues like "spiritual enligthenment" (to reduce urban existential suffering) and "social justice" (for the poor, minorities, and other unfortunate ones in society)—as will be clear in the next section and the rest of the book.

Unfortunately, not every scholar in urban planning agrees with the policy implications of the main features of sustainability.

So, the question here is, Why?

# The Theoretical Debate

To answer this question requires an introduction to the theoretical debate on the issue of sustainability in urban planning (and other related fields).

Four main theoretical approaches can be summarized hereafter (and also in *Table 1.2*), with the fourth to be my original contribution to the literature (as summarized in Chapter Four, the concluding chapter).

These four theoretical approaches can be called, in the absence of better terms, namely, (a) the *engineering* argument, (b) the *ecology* argument, (d) the *fairness* argument, (d) the *complexity* argument, and (e) the *contingency* argument—to be analyzed hereafter, in that order.

## The Engineering Argument

Firstly, a major theoretical approach can be called, in the absence of better words, as the *engineering* argument, which advocates a top-down approach towards urban planning, especially though not exclusively with the use of technology.

The engineering argument refers to the "systems engineering" tradition which can be traced back to the "planning beginning with Saint Simon and August Comte in the early 1800s. There is a pervasive, 'systems thinking' influence throughout the history of planning that is based on the engineering control paradigm." (M. Dempster 1998)

In the later part of the $20^{th}$ century, "addition of systems analysis to the rational model…has created the sense of a more exact, scientific approach….This particular approach has been integral to planning, especially within institutional frameworks, though to a substantially lesser degree in radical forms of planning." (M. Dempster 1998; J. Wolfe 1989)

A good illustration concerns the "strategic urban planning" (SUP) which suggests "clarifying which city model is desired and working towards that goal, coordinating public and private efforts, channelling energy, adapting to new circumstances and improving the living conditions of the citizens affected." (WK 2008j)

For instance, the city of San Francisco in California "carried out…[the] process [of SUP] between 1982 and 1984. The main motivation behind starting strategic urban planning processes was the attempt to adequately react to problematic situations (mainly economic crisis or standstill)." (WK 2008j)

For the critics, however, the top-down approach in the engineering argument is elitist, in that it does not pay sufficient attention to "civic participation" and "political activism" for the less well offs in the city. (WK 2008j)

Another serious problem with the "engineering control paradigm" for urban planning is the inherent "difficulties of societal guidance," as John Friedmann (1987 & 1996) thus said: "Because it is invariably integrated into the state apparatus, planning for societal guidance is incapable of coping with the crisis of industrial capitalism." (M. Dempster 1998)

## The Ecology Argument

Secondly, another theoretical approach can be referred to, in the absence of better terms, as the *ecology* argument, in that, unlike the engineering argument, it pays more focus on the close linkages between the ecological system and the human environment in urban planning.

In fact, the ecology argument rejects the greedy presumption of "profit maximization" so much championed in lasses-faire capitalism (as famously shown in the magic of the "invisible hand" by Adam Smith and his successors in the neo-classical economic tradition), which often ignores or does not pay serious attention to the problem of ecological destruction in the classical developmental process, which is thus criticized as un-sustainable.

As M. Beth Dempster strongly argued, "the neoclassical economic interpretation…typically fails to recognize our dependence on environmental inputs….This is particularly important when it is noted that societies that perceive their resources as infinite are not likely to reduce their impacts….This is especially problematic when it is recognized that a society can misinterpret the state of their resources." (M. Dempster 1998; M. Gadgil 1991; E. James 1991)

This problem of unsustainability is all the more serious, when it has now reached the global scale of greenhouse effects, to the point that "we are talking about our planetary survival system." (M. Dempster 1998; K. McNamee 1996)

In other words, "our impacts on biophysical systems have achieved a fundamental change in scale both spatially and temporally. Humans are influencing biophysical systems on a global scale and have set a future trajectory of impacts for a questionable time into the future as illustrated by concerns regarding destruction of the ozone layer and climate change issues. Our conventional institutional arrangements are not necessarily able to cope with the changing circumstances." (M. Dempster 1998; N. Doubleday 1993)

Yet, for the critics, the ecology argument is flawed, not because it is not correct in pointing out the serious ecological problem challenging "our planetary survival system," but because the nature of sustainability is complex and thus uncertain in outcome.

To put it differently, "sustainability can only be determined in hindsight," because "we operate within the paradox of interdependence, requiring a predictable environment which, to retain its adaptive ability, must continually change." (M. Dempster 1998; R. Costanza 1996)

So, the ecology argument is naïve, since "[w]e can never know if current activities are sustainable, but can only infer their potential….These are complicated by recognition of the complex, unpredictable systems characteristics" in nature. (M. Dempster 1998)

## The Fairness Argument

Thirdly, a different (though related) theoretical approach can be called, in the absence of better terms, the *fairness* argument, in that any urban planning should consider the issue of social justice, especially for those who are disadvantaged, be they the poor, ethnic/racial minorities, social deviants, and others, who often have the misfortune to live in bad urban neighborhoods, while the rich enjoy nicer ones.

For instance, Jason Corburn (2008) argued for "an environmental justice framework that connects ecological, economic and social justice issues, especially in urban settings. Corburn asks about the distribution of environmental goods…, who participates in rule-making and enforcement, and how environmental justice evolves institutionally, and is enforced."

Phillip Thompson (2008), for another, spoke "for many lower-income city dwellers" when he asked, "Who, for example, will pay for outfitting buildings in poorer neighborhoods…and who will train and educate the workforce required for this transformation?"

Chris Zegras (2008) likewsie championed the need for the alleviation of the "fundamental issues of accessibility for the poor: their lack of affordable transportation and proximity to schools and jobs. Zegras recommends addressing the worldwide crisis in transportation, in part through such innovations as bike and car sharing."

And Adil Najam (2008) shared the same argument by observing that, if the earth is one single country, then it "is poor, extremely divided, degraded, poorly governed and unsafe—a Third-world country."

Yet, for the critics, the intention of all these scholars is noble indeed, fighting for social justice on behalf of the poor, the minorities, and other unfortunate ones.

But, the deeper question for them here concerns the age-old question which has divided philosophers through the ages, namely, What is justice? Who is going to say what is just or unjust, fair or unfair, so that every group on earth can agree?

In *The Future of Human Civilization* (*FHC*), I already argued why this search for social justice on earth is essentially conflicting, to the extent that there is no morality without immorality, no justice without injustice, or no equality without inequality.

## The Complexity Argument

Fourthly, an alternative theoretical approach can be referred to as the *complexity* argument, in that it takes seriously the "complexity" of the closed linkages between the natural environment and the human one in urban planning, in a way which tries to resolve the problem of uncertainty in outcome—unlike the ecological argument, which is rejected (by those in the complexity argument) as a bit too naïve, albeit noble in intention.

The understanding of "complexity" is not an easy task. For instance, some scholars like Seth Miller go so far as to propose "at least 45 different definitions of 'complexity.'" (J. Rosser 2003) Others, however, like David Green (2001) and David Newth, identify six main features of complexity, namely, (i) "self-organization," (ii) "connectivity," (iii) "criticality," (iv) "novelty," (v) "diversity," and (vi) "emergence"—as this is something that I already went to great lengths to analyze in *The Future of Complexity* (*FC*).

With this difficulty in mind—it is not surprising that, while "it is critical to recognize fundamental differences between the 'engineering' paradigm and newer approaches based on complex systems concepts," the "[a]pplication of 'new' systems thinking (e.g. self-organization, complexity) to planning seems to be quite limited as yet in the formal literature." (M. Dempster 1998)

Even then, "[d]espite this lack of application of new systems thinking, the trend toward alternative planning approaches suggests a growing recognition of planning's sympoietic nature....The flexible, adaptable, communicative, transactive, collaborative characteristics [were] espoused by a host of planners." (M. Dempster 1998).

Good examples include the works of L. Susskind (1985), S. McCreary, J. Forester (1989), M. Moore (1991), J. Friedmann (1993), J. Nelson, R. Serafin (1996), S. Mandlebaum (1996), J. Nelson (1997), A. Skibicki, and others.

But the critics are quick to point out two essential limits in the complexity argument for resolving the issue of sustainability in urban planning.

Firstly, the first problem concerns "our current abilities and approaches. Can we cope with the complexities and uncertainties inherent in these systems? In particular, can we cope with the sympoietic systems that form the environment upon which we depend? Do we recognize the differences between the two systems types and treat them appropriately? Do we recognize the implications of structural coupling and organizational closure? Do we have planning approaches that carry sympoietic characteristics? Do we recognize the implications of future causality? Does planning play an adequate role as an integral part of our complex, adaptive, sympoietic social system?" (M. Dempster 1998)

M. Beth Dempster (1998) thus confessed: "My general answer to these questions is—not adequately."

And secondly, the second problem is a follow-up to the first, in that if we now understand the uncertain nature of many complex systems, the challenge then is to come up with a way to address the issue of sustainability within a complex system more realistically.

As Dempster (1998) asked, "How can and should we cope with these system complexities? What planning approaches should be encouraged or developed to ensure appropriate treatment of autopoietic and sympoietic systems? What role should planning play in our social systems? Can planning help to make these systems adaptable? Should we generate sympoietic planning approaches, and if so, how?"

Again, Dempster's answer is also uncertain: "In general, I believe planning has potential to act as a key contributor to human adaptability. To play this role, however, planning must be able to understand and cope with complex natural and social systems and their interaction. Planning for sustainability must be capable of coping with the interaction between 'causes' of human behaviors and their actualization,...and structural coupling of social systems to the environments they are embedded within. It must be structured to allow coping with the uncertainties and complexities inherent in both natural and social sympoietic systems. It is imperative to recognize the recursivity and self-organization of poietic systems and to acknowledge the disadvantages, but to capitalize on the advantages." (M. Dempster 1998)

But, for the critics, the challenge here remains: How exactly should urban planning be done in such a way as to deal with "the paradox of interdependence, requiring a predictable environment which, to retain its adaptive ability, must continually change," together with its uncertain characteristics in outcome which are typical in complex systems?

### The Contingency Argument

And fifthly, unlike the previous theoretical approaches—an alternative one can be called, in the absence of better words, the *contingency* argument, which learns from the previous four theoretical approaches without, however, favoring one argument over the other and, in the process, transcends them all.

The contingency argument, in the end, has the virtue to show that the concern with sustainability has been much exaggerated and distorted, to the point that it is fast becoming a new intellectual fad, so that its dark sides have been unwarrantedly ignored or downgraded.

Thus, the contingency argument is my original theoretical contribution to the debate on the issue of sustainability in the context of urban planning, with the analysis of "density" and "void" as a case study.

## The Contingent Theory of Urban Planning

My *contingency* argument can be more precisely called *the contingency theory of urban planning*, since it proposes a new theory of urban planning for the study of sustainability.

This contingency theory of urban planning transcends the previous four arguments by adding something new to the literature, yet without indulging in the euphoric discourse on the possibility and desirability of sustainability.

In other words, the contingency theory of urban planning has seven major theses, namely, (a) the first thesis: the denseness-emptiness principle, (b) the second thesis: the absoluteness-relativeness principle, (c) the third thesis: the simpleness-complicatedness principle, (d) the fourth thesis: the evolution-transformation principle, (e) the fifth thesis: the symmetry-asymmetry principle, (f) the sixth thesis: the regression-progression principle, and (g) the seventh thesis: the post-human planning—to be elaborated in the rest of the book, with a summary in the concluding chapter.

It is important to note, however, that other principles (besides the six as cited above) are also relevant, but these six are the most relevant in the current case study (as will be explained further in Chapter Four, the concluding chapter).

This qualification aside—this seminal theory, if true, will fundamentally change the way that we think about the issue of sustainability by transcending it, with enormous implications not only for understanding the future of urban planning, in a small sense—but also for predicting the relevance of sustainability in relation to the entire domain of human knowledge for the human future and what I originally called its "post-human" fate, in a broad sense.

## Theory and Meta-Theory

The contingent argument, or the contingency theory of urban planning, presupposes some kind of meta-theory, in special relation to methodology and ontology—just as all theories, either explicitly or implicitly, require their own versions of meta-theory.

In my case, I already proposed, in my previous works, my distinctive approach to methodology (known as "sophisticated methodological holism") and ontology (known as "existential dialectics").

The next two sections are to introduce my meta-theory, or existential dialectics and sophisticated methodological holism, respectively.

## The Logic of Existential Dialectics

To start, the ontology of existential dialectics can be summarized in this section—to be repeated, *often verbatim* (with only a few updated revisions), from my previous works—as this is something that I regularly do in each book as an introduction to some background information for the convenience of the readers.

After all, this book, like all others of mine, is in conversation with all previous ones of mine, for the final aim to converge all of them into a larger grand project on the future of intelligent life, both here on earth and there in deep space unto multiverses.

The summary can be organized in relation to (a) the conception of existential dialectics (or its ontics), (b) the syntax of existential dialectics (or its ontomethodology and ontologic), (c) the semantics of existential dialectics (or its ontosemantics), and (d) the pragmatics of existential dialectics (or its ontopragmatics), respectively hereafter—again *often verbatim*.

This means that, for those readers who had read some of my previous books, the summary below is really nothing new (*almost verbatim*, with only some minor revisions) and can be skipped.

But for those who have never read my previous books, it is a good review of them.

## The Conception of Existential Dialectics (or Its Ontics)

This ontology was originally designated as "existential dialectics" in *Beyond Capitalism to Post-Capitalism* (herein abbreviated as *BCPC*), although it was already analyzed in other books of mine like the 2 volumes of *The Future of Human Civilization* (*FHC*), *The Future of Capitalism and Democracy* (*FCD*), *The Future of Post-Human Consciousness* (*FPHC*), and the 2 volumes of *Beyond Democracy to Post-Democracy* (*BDPD*).

Lest any confusion occurs, it is important to stress at the outset that the word 'existential' in "existential dialectics" has nothing to do with Existentialism, which I rebuked in *FHC*, *FCD*, and also *FPHC*. Instead, the word merely refers to the existence of intelligent life (both primitive and advanced) in a broad sense.

*Some Basic Concepts*

The conception of existential dialectics (or its ontics) makes use of different concepts (like "sets," "elements," "relations," "operations," "functions," "truth values," "axioms," "postulates," and "principles"—as shown in *Table 4.3*), which are important for the understanding of any logic of ontology.

That clarified—existential dialectics, as a language of ontology, can be analyzed in different ways that a language, as an analogy, is often analyzed, namely, in relation to phonology ("the study of patterns of a language's sounds"), phonetics ("the study of the physical aspects of sounds of human language"), morphonology ("the study of the internal structure of words"), syntax ("the study of how words combine to form grammatical sentences"), semantics ("the study of the meaning of words [lexical semantics] and fixed word combinations [phraseology]"), and pragmatics ("the study of how utterances are used…literally, figuratively, or otherwise…in communicative acts"). (WK 2007)

But since my theory of existential dialectics makes use of the English language for communication here, it does not propose a new way to make the patterns of a language's sounds (as in phonology), to study the physical aspects of sounds (as in phonetics), or even to strive for a new organization of the internal structure of words for the English language (as in morphonology).

For this reason, the language of existential dialectics to be analyzed hereafter is not concerned with the phonological, phonetic, and morphonological aspects of the English language. Instead, the analysis will explore the syntax, semantics, and pragmatics of existential dialectics as a language of ontology to improve our understanding of reality.

*The Ontic Dispute*

But first, there is an ontic dispute to be addressed, which is related to what I called "the dilemma of ontology" as introduced earlier in *Sec. 1.4* of *The Future of Post-Human Knowledge* (*FPHK*).

A good point of departure concerns what I want to call *the ontic dispute*, on how to understand the very nature of ontology.

Traditionally, scholars in the field used to study the conception of ontology in terms of its entities (e.g., material objects, abstract numbers), properties (e.g., duration, plurality), and relationships (e.g., causation).

This is a good starting point, but it suffers from different problems. Consider three examples of bias below, which are quite common in the literature, and the recent version (of treating concepts and principles in ontology on the basis of "plurality," "dynamism," "duration," "interaction," "life," "consciousness," and "volition") by Reginald Firehammer (2005) only constitutes a latest addition to this old (impoverishing) tradition.

Firstly, the classification of entities, properties, and relationships is too material-centric, in often focusing more on the existence of material entities than on non-material ones.

Secondly, the classification is too anthropo-centric, in often favoring the factor of human agency in depicting reality.

And lastly, the classification is too confused among the three categories relating to each other, in often lumping them together without due consideration of their intricated interrelationships (both in relation to each and in relation to all).

A good alternative is precisely what I want to offer in this debate, that is, a new classification of ontological entities, properties, and their relationships in terms of *method*, *structure*, *process*, *agency*, and *outcome* instead.

In relation to method, it concerns how ontology is to be studied. In this sense, the word 'method' here should not be confused with the different usage of the same term 'method' as a technique of investigation in the context of methodology (as already introduced in *Sec. 1.2* of *FPHK*).

In relation to structure, it concerns what the general patterns of entities, properties, and relationships are.

In relation to process, it concerns how these general patterns change over time.

In relation to agency, it concerns how agency affects any change of these general patterns over time (without putting, however, too much emphasis on the role of human agency).

And in relation to outcome, it concerns what the outputs of this change over time are.

## The Syntax of Existential Dialectics (or Its Ontomethodology and Ontologic)

The syntax of existential dialectics, analogously speaking, refers here to the structure of an ontology which can be studied by way of combining ideas into complex relationships like ontological principles to depict reality in the world—and can be analyzed in seven sub-sections, namely, (a) the selection criteria, (b) further clarifications on classification, (c) ontomethodology, (d) ontologic, (e) the principles as short cuts, (f) the principles as family resemblances, and (g) the dialectic constraints imposed by the principles, respectively hereafter.

### *The Selection Criteria*

To start, not any pair of relationship can be chosen as an ontological principle in existential dialectics; otherwise, there could be billions of them under the sky.

Four illustrations below suffice to clarify this point.

(a) Firstly, an ontological principle cannot *overlap* with other principles, in that it is too closely related with other ones.

Of course, one cannot totally rule out any relationship between two entities, especially in a complex world of everything being linked to everything else—but the comparison here is relative, not absolute.

In this sense, the flexibility-inflexibility pair cannot be an ontological principle, since it has something closely in common with other ones (like the simpleness-complicatedness principle, for instance).

As an illustration, something which is flexible in interpretation is more likely to allow multiple viewpoints and thus makes the overall picture more complicated than otherwise (simple).

(b) Secondly, an ontological principle cannot be *redundant* in relation to other principles, in that it somehow duplicates other principles.

In this sense, for instance, the directness-indirectness pair cannot be an ontological principle, because it somehow duplicates other ontological principles (like the preciseness-vagueness principle, for instance).

As an illustration, something which is direct means to be right to the point and does not need to go to an unnecessarily long loop—so it is less vague, in being clearer about the thing in quesiton.

(c) Thirdly, an ontological principle cannot be *derived* from other ones—as if it is a child-parent relationship.

In this sense, the convenience-inconvenience pair cannot be an ontological principle, because it can be derived from other ones (like the slowness-quickness principle, for instance).

For instance, something which is convenient already implies that it can be obtained quickly (rather than slowly)—all things considered.

(d) Fourthly, an ontological principle cannot be *trivial* (or *parochial*)—as if it is only one component competing with other ones for inclusion in a set.

In this sense, the consistency-inconsistency pair cannot be an ontological principle, because its validity is limited (or parochial), as it can be easily challenged and replaced by opposing major (not minor) theories (e.g., the correspondence theory of truth, the pragmatic theory of truth, the semantic theory of truth, and whatnot), not just in accordance to the consistency theory of truth (for consistency).

In this sense, the consistency argument is only one among different opposing arguments in relation to the logic of what constitutes "truth."

Although these examples are not exhaustive, they illustrative the selectiveness of any ontological principle to be existentially dialectic.

### *Further Clarifications on Classification*

Lest any misunderstanding carelessly occurs, some further clarifications on classification should be made here (and summarized in *Table 4.8*).

(a) Firstly, the total number of ontological principles is unknown, to be discovered later, as our knowledge of the world becomes more advanced.

Consequently, the principles as introduced in my books are not exhaustive, with new ones being added, whenever more of them are discovered in later research.

At least, future generations can pick up where I leave off and continue the discovery.

In this sense, my work should be treated as a pioneering effort for the development of a systematic, comprehensive analysis of a new general ontology for the future of knowledge.

(b) Secondly, the principles are not rigidly classified, as they can be reclassified in a different way.

For instance, the preciseness-vagueness principle is classified under the category of "structure" but can be reclassified under the category of "method," although in so doing, it has a different meaning in the context of method.

The same logic applies to the same-difference principle under the category of "outcome," which can be reclassified under the category of "structure," although in so doing, once more, it has a different meaning in the context of structure.

(c) Thirdly, the principles are to be understood together, not that each prinicple is to be treated independently of others. After all, the principles are mutually constraining, in that they work together as a whole.`

(d) Fourthly, the principles are relevant to all subject matters, but some principles are more useful to some subject matters than others—as implied in the symmetry-asymmetry principle.

And (e) fifthly, the principles constitute only two levels of analysis, this time, at the ontological and methodological levels—while other levels of analysis (from the perspectives of the mind, nature, society, and culture) are also needed, in order to understand reality in its totality (as explained in the section on "sophisticated methodological holism," so as to avoid the dual dangers of reductionism and reverse-reductionism).

*Ontomethodology*

With these selection criteria and clarifications in mind—the syntax of existential dialectics can be divided into two main smaller sections for analysis, since they are related to each other in the context of some fundamental principles, that is, (a) ontomethodology and (b) ontologic, to be addressed hereafter, respectively.

In the context of ontomethodology, some ontological principles are proposed here in relation to the nature of methodology, which however is tied up with the general issue of ontology in the context of knowledge.

In this sense, ontomethodology (in the context of these ontological principles in the category of method) are related with ontologic in general (in the context of ontological principles in general).

Some good instances of ontological principles involving ontomethodology include the formalness-informalness principle, the predictability-unpredictability principle, the explicability-inexplicability principle, the partiality-totality principle, and the absoluteness-relativeness principle, which will be introduced in conjunction with ontological principles in the ontologic of existential dialectics in general hereafter.

*Ontologic*

With this clarification of the relationship between ontomethodology and ontologic in mind (which is comparable to the relationship between methodology and ontology as described in *Sec. 1.3* of *FPHK*)—the first three principles in the ontological logic of existential dialectics were introduced in *BCPC*. And more principles were later added to the logic in my subsequent books (like *FC*, *FAE*, and others).

More specifically, in *BCPC*, I proposed three principles (based on previous works of mine, not just something out of the blue) for the ontological logic, namely, (a) the regression-progression principle on the direction of history, (b) the symmetry-asymmetry principle on the relationships among existents, and (c) the change-constancy principle on the alteration of things—or in a more elegant term, the dynamics of space-time (in *FC*).

In *The Future of Complexity* (*FC*), I added three more principles, on top of the three aforementioned, namely, (d) the partiality-totality principle on the relationships between the parts and the whole, (e) the order-chaos principle on the pattern of things, and (f) the predictability-unpredictability principle on the occurrence of things—as part of the ontological logic of existential dialectics.

In *The Future of Aesthetic Experience* (or simply *FAE*), another principle is added, namely, (g) the evolution-transformation principle (on the multiple kinds of agency).

Also in *FAE*, I then reclassified the 7 principles into four categories, namely, (i) in relation to method, for the partiality-totality principle and the predictability-unpredictability principle, (ii) in relation to process, for the change-constancy principle and the order-chaos principle, (iii) in relation to agency, for the symmetry-asymmetry principle and the evolution-transformation principle, and (iv) in relation to outcome, for the regression-progression principle.

In *The Rise of Authoritarian Liberal Democracy* (*ALD*), I added two more principles, namely, (h) the softness-hardness principle on the force of agency (which is to be classified under the category about agency) and (i) the same-difference principle about the metamorphosis of change (which is to be classified under the category about outcome).

In *The Future of Information Architecture* (*FIA*), I introduce three additional principles, that is, (j) the simpleness-complicadness principle on the interconnection among things, (k) the preciseness-vagueness principle on the refinement of things, and (l) the slowness-quickness principle on the speed of change—with the first two in relation to structure and the third in relation to process.

In *The Future of Post-Human Unconsciousness* (*FPHU*), I added three more principles, that is, (m) the openness-hidden principle on the detection of things, in relation to structure, (n) the explicability-inexplicability principle on the underlying mechanisms of things, in relation to method, and (o) the expansion-contraction principle on the growth of things, in relation to process.

In *The Future of Post-Human Knowledge* (*FPHK*), I added one more principle, that is, (p) the absoluteness-relativeness principle on the multiplicity of entities, in relation to method.

In *The Future of Post-Human Mathematical Logic* (*FPHML*), I added one more principle, that is, (q) the formalness-informalness principle on the formal requirements of systems, in relation to method.

In *The Future of Post-Human Engineering* (*FPHE*), I added one more principle, that is, (r) the theory-praxis principle on the duality of knowledge, in relation to agency.

In *The Future of Post-Human Creative Thinking* (*FPHCT*), I added one more principle, that is, (s) the convention-novelty principle on the nature of creative thinking, in relation to agency.

In *The Future of Post-Human Geometry* (*FPHG*) here, I added one more principle, that is, (t) the finiteness-transfiniteness principle on the nature of numbers, in relation to structure.

And in *The Future of Post-Human Urban Planning* (*FPHUP*) here, I add one more principle, that is, (u) the denseness-emptiness principle on the distribution of entities in space, in relation to structure.

With this update, there are twenty-one principles so far in existential dialectics, that is, five principles for the category in relation to method (viz., the formalness-informalness principle, the partiality-totality principle, the predictability-unpredictability principle, the explicability-inexplicability principle, and the absoluteness-relativeness principle), five principles for the category in relation to structure (viz., the finiteness-transfiniteness principle, the simpleness-complicatedness principle, the preciseness-vagueness principle, the openness-hiddenness principle, and the denseness-emptiness principle), four principles for the category in relation to process (viz., the change-constancy principle, the slowness-quickness principle, the order-chaos principle, and the expansion-contraction principle), five principles for the category in relation to agency (viz., the theory-praxis principle, the convention-novelty principle, the symmetry-asymmetry principle, the evolution-transformation principle, and the softness-hardness principle), and two principles for the category in relation to outcome (viz., the regression-progression principle and the sameness-difference principle).

There are thus twenty-one principles in existential dialectics—so far (as summarized in *Table 4.4*).

*The Principles as Short Cuts*

Yet, these principles should be treated with caution, lest misunderstanding occurs, since they do not constitute rigid dualities (or dichotomies).

The reason is that each pair in an ontological principle consists of two opposites, which are, however, merely short cuts both for multiple variations and degrees, as well as for different interactions with multiple other entities unlike them.

In this light, each pair can end up having hundred (or even thousand, if not more) different versions, which interact with hundred (or even thousand, if not more) other entities—as summarized in *Table 4.5*.

There are two clarifications here.

Firstly, this conception of shortcuts is not mutually exclusive nor absolute, in that the opposites can come in all shapes and sizes, with different degrees. For instance, by analogy, just as there are different degrees of the two colors "white" and "black"—there are likewise different degrees of the opposites in each ontological pair, to the extent that there can be multiple entities (not only two) interacting with each other in each pair.

And secondly, this conception of shortcuts do not ignore other possible entities in interacting with the two opposites (with their different versions) in each pair. For instance, by analogy, there are not only the two colors "white" and "black" as opposites, since there are other colors too besides them like "yellow," "green," "purple," or else. The same logic can be applied to each ontological pair, in that they also interact with other entities, not with only two of them (with their different versions).

Therefore, with these two clarifications in mind—each pair in an ontological principle serves only as an abbreviation for something more complicated and, therefore, although it contains two opposites, it should not be confused as a duality (dualism).

Existential dialectics rejects any dualism (or dichotomy) as too rigid and instead allows the multiplicity of entities, to the extent that between the two opposites in each pair exist many other alternatives to choose from. They are named in that short form for aesthetic elegance, instead of listing all possible entities between the two opposites in the title.

To be dialectic is to go beyond any rigid dichotomy and transcend into something different altogether in the long haul. One may be tempted to call the dialectic logic here with a different name like existential "multilectics" (instead of "dialectics"), but this naming is incorrect (or even misleading), for the two reasons aforestated.

After all, each ontological pair come in all shapes and sizes (with different degrees) and do not exist by themselves but also interact with other entities unlike them. The virtue of revealing an ontological pair is to show how they relate within themselves (in multiple versions) and also interact with others unlike them (also in multiple versions).

This will be clear in a later sub-section titled "the dialectic constraints imposed by the principles."

*The Principles as Family Resemblances*

In addition, each ontological principle is generic, with some other comparable ontological pairs to be put in the same family (like a *family resemblance*)—as summarized in *Table 4.6*.

Ludwig Wittgenstein (1953) once suggested the idea of "family resemblance" in explaining different games classified under the same family called *games*.

Why should, for instance, playing football and chess as playing "games," when it is well understood that football is not the same as chess? The answer is that, although each game is different and has different rules, many of them (though not all) share, more or less, some commonalites (e.g., scoring as necessary for winning).

And this is so, even though some games share more than some others in any given selection of critieria, and no two games are exactly identical. So, Wittgenstein's point here is that there is no essential core which is common to all games, and the best that one can look for is some characteristics which are common to many (but not all) games. (A. Biletzki 2006)

By the same logic—in the previous section on selection criteria, the flexibility-inflexibility pair can be put in the family resemblance of the simpleness-complicatedness principle, although the two pairs are not exactly identical. Likewise, the directness-indirectness pair can be put in the family resemblance of the preciseness-vagueness principle, although, again, the two pairs are not exactly identical.

In this sense, which specific pair in a family should be used to designate the name of the family can be at times a bit arbitrary, but with good reason.

For illustration, in the context of method, the partiality-totality principle can take the different form like individualisticness-holisticness, just as the explicability-inexplicability principle can take the different form like underlyingness-regularness—although each two pairs are not exactly identical and have slightly different meanings and usages.

A more comprehensive listing of this family resemblance for all other principles is shown in *Table 4.4* on the syntax of existential dialectics.

In fact, this section on family resemblances overlaps with the semantics of existential dialectics, which will be introduced later in the next section (and summarized in *Table 4.10*).

*The Dialectic Constraints Imposed by the Principles*

The principles, as they constitute the syntax of existential dialectics, are dialectic in character, such that, when they are applied, they impose dialectic constraints on how reality is to be understood.

Consider, say, the symmetry-asymmetry principle as an illustration here, in order to summarize two main characters of the dialectic constraints in question (as summarized in *Table 4.7*).

Firstly, to be dialectic here is to go beyond the narrow dichotomies (and, for that matter, any rigid multi-dimensional classificatory scheme), be they about "self" vs. "world," "freedom" vs. "unfreedom," "barbarity" vs. "civilization," "individuality" vs. "communality," and so on.

One way to do so (to go beyond) is to consider them all in terms of co-existence (without favoring one over the rest). For instance, my theory of "post-civilization" (to be summarized later in the section on the pragmatics of existential dialectics) is to go beyond barbarity and civilization in terms of understanding barbarity and civilization as being co-existent.

And the same logic can be said in relation to my theories of "post-democracy," "post-capitalsm," and others (also to be introduced later in the section on the pragmatics of existential dialectics), in regard to freedom vs. unfreedom, equality vs. inequality, communality vs. individuality, spirituality vs. materiality, and so on.

But to consider them all (in the dichotomies—and, for that matter, in any rigid multi-dimensional classifactory schme) as co-existent is not the same as to imply that the opposing categories in any classificatory scheme are all equal, since, in accordance to the symmetry-asymmetry principle (as an illustration here), if they are equal in terms of being considered as co-existent, they are asymmetric in terms of being unequal in dialectic interaction (e.g., X can be more dominant than Y in case A, or Y is more dominant than X in case B), although in some special cases, they can be relatively equal.

For this reason, there are different versions of "post-democracy" and "post-capitalism" in my theories. As an illustration, in version I of the theory of post-democracy, freedom is more dominant than equality, whereas in version II of the theory of post-democracy, equality is more so than freedom.

But this "X more than Y" has to be understood in the context of dialectic logic (not in symbolic logic, as conventionally understood), in that both "X" and "Y" are important in post-democracy (in the context of dialectic logic), but in an asymmetry way.

By contrast, in symbolic logic, it often favors one over the other—be it in regard to privileging freedom over equality in Fascism, favoring freedom relatively more than equality in Liberal Democracy, or favoring equality relatively more than freedom in Socialist Democracy. In regard to the latter two cases (about Liberal Democracy and Socialist Democracy), the difference between dialectic logic and symbolic logic can be one in degree, not in kind—in this sense, albeit not in other senses.

The same reasoning can be said about the relatonships between individuality and communality, between spirituality and materiality, and between formal legalism and informal legalism in the different versions of my theory of "post-capitalism."

Secondly, to be dialectic is to go beyond the narrow dichotomies (and, for that matter, any rigid multi-dimensional classificatory scheme) in another way, this time, in a transcendent way, that is, in exploring other possibilities or even other issues not considered within the narrow confines of narrow dichotomies (and, for that matter, any rigid multi-dimensional classificatory scheme).

As an analogy, in this second meaning, to go beyond the narrow dichotomy of "black" vs. "white" is not just to choose both "black" and "white" (as in the first meaning) but also to explore other options (e.g., "green," "purple," "blue," etc.—and, alternatively, "shade," "line," "curve," etc.).

By the same reasoning, to go beyond "democracy" is to transcend democracy (as in version III of my theory of "post-democracy") and to explore other possibilties of lifeforms (e.g., "floating consciousness," "hyper-spatial consciousness," etc., to live beyond the narrow obsession with freedom and equality).

This dialectic character of the principles in existential dialectics has important implications for the pragmatics of existential dialectics (as will be clear shortly, in the section on the pragmatics of existential dialectics).

## The Semantics of Existential Dialectics (or Its Ontosemantics)

The syntax of existential dialectics so understood in terms of ontological principles only gives us the structure of ontology in the world, in an abstract (general) sense. These principles by themselves do not tell us the specific meanings in a given context.

In order to grasp the specific meanings of the principles in a given context, it is necessary to study the semantics of existential dialectics (as summarized in *Table 4.10*).

And the analysis of the ontological principles as family resemblances in the previous section on syntax is only a starting point (and thus overlaps a bit with the semantics of existential dialectics here).

For this reason, I have often gone to great lengths in my previous books on different subjects to explain the specific meanings of the principles when applied in different contexts.

Consider two examples below for illustration, which are not exhaustive.

In *FPHST*, for instance, I used the first three principles (i.e., the change-constancy principle, the regression-progression principle, and the symmetry-asymmetry principle) to propose "the perspectival theory of space-time," for a better way to understand space and time—especially, though not exclusively, in relation to future post-human history (as summarized in *Table 1.32*, *Table 1.33*, *Table 1.34*, and *Table 1.35*).

In so doing, I had to introduce concepts and theories specific to the field of physics and other related fields (e.g., "absolute space" and "absolute time" in "classical mechanics" and "relative space-time" in "the theory of relativity").

And in *BNN*, I also exploited the three principles to propose the "transcendent" approach to the study of genes and memes as a new way to understand the interaction between nature and nurture.

In so doing, I had to explore concepts and theories in the world of evolutionary biology (e.g., "mutation," "variation," "adaptation," "selection," and "inheritance" in classical evolutionary theory) and neural biology (e.g., "chromosome," "gene," "DNA," "RNA," "protein," "neuron," "neural network," and "behavior").

Other examples, besdies these two here, are given in other books of mine on different topics.

Therefore, this semantic dimension of existential dialectics is important, since, in each of my books when the ontological principles are used, each principle becomes relevant in a different meaning when contextualized within a given case study in question.

In this book, for instance, the principles are relevant in relation to the subject matter of urban planning (as summarized in the concluding chapter), and they can be shown in all shapes and sizes (since they are also to be treated as "short cuts," as indicated before), without, however, being reduced to "specific ontology" (as opposed to "general ontology"), as explained in *Table 4.9*.

## The Pragmatics of Existential Dialectics (or Its Ontopragmatics)

Even the semantics of existential dialectics is not enough to understand reality, since semantics by itself does not tell us the nature of intentionality in relation to the matrix of power formations and human interests when existential dialectics as an ontology is applied.

Surely, one cannot completely separate the pragmatics of existential dialectics from its semantics or syntax, but the point here is to give each a more elaborated analysis.

For this reason, the pragmatics of existential dialectics needs to be studied too, in special relation to the inclusion of power formation and human interests in the application of the principles of existential dialectics (as summarized in *Table 4.11*).

It is thus no wonder that all my previous books have shown in different ways how and why these ontological principles and their theoretical applications can reveal the future world to come in a way that humans have never known, in a totally different battleground for competing human (and later post-human) interests by myriad groups fighting for their own versions of hegemony.

Perhaps a summary of my previous works in relation to existential dialectics and their contributions to my new theories as proposed over the years is deemed revealing here, for the further understanding of the pragmatics of existential dialectics (in close relation to the semantics of existential dialectics too).

That stressed—the following summary in the next five sub-sections is something that I often do in my previous books too, as a way to introduce the intellectual background of my ideas.

So, for those readers who had read my previous works, the following sub-sections serve as a reminder, as they are repeated hereafter *often verbatim* (with only some minor changes, with the rest kept intact).

These sub-sections are, namely, (a) the two-way street connecting theory with meta-theory, (b) the theoretical application of existential dialectics, (c) the further use of existential dialectics, (d) direct and indirect applications of existential dialectics, and (e) multiple applications of existential dialectics—to be summaried herefafter, respectively.

*The Two-Way Street Connecting Theory and Meta-Theory*

However, lest any misunderstanding occurs, it is important to remember that the pragmatics of existential dialectics is not a one-way street (that is, using the ontological principles for theoretical insights in praxis) but a two-way one, namely, (i) from meta-theory to theory, and (ii) from theory to meta-theory.

(i) On one side of the street, the ontological principles can inspire some theoretical insights in praxis, that is, in relation to some specific fields.

(ii) On the other (opposing) side of the street, however, the study of a subject matter in the specific fields in turn reveals some more hitherto unknown ontological principles to be discovered and identified.

For this reason, three new principles were added in *FC*, one in *FAE*, two in *ALD*, and three in *FIA*, on top of the original three in *BCPC*, for example (as there are more)—after some research on the specific subject matters.

Both theory and meta-theory enrich each other in all my works over the years.

*The Theoretical Application of Existential Dialectics*

For instance, at the beginning, in *FHC*, two examples of the theoretical application of existential dialectics are to reveal no freedom without unfreedom (as shown in *Table 4.12*) and no equality without inequality (as shown in *Table 4.13*), especially in relation to the seven dimensions of life existence first worked out in *FHC* (i.e., the technological, the everyday, the true, the holy, the sublime/beautiful, the good, and the just).

In Ch.10 of *FCD*, the relationship between these two examples of the theoretical application of existential dialectics was extended further, so as to provide some new visions to understand the future forms of political and economic systems.

In regard to politics, my vision of a path-breaking political system in future times concerns the different forms of "post-democracy" to supersede democracy unto the post-human age.

My vision of post-democracy was called "the theory of post-democracy" (as summarized in *Table 4.20*, *Table 4.21*, and *Table 4.22*), whereas *Table 4.23* explains the distinctions among democracy, non-democracy, and post-democracy.

And *Table 4.24* stresses the multiple causes of the emergence of post-democracy, while *Table 4.25* clarifies some possible misunderstandings in regard to post-capitalism and post-democracy.

In *ALD*, I also showed how authoritarianism can be an advanced form of liberal democracy (just as it has its own counterpart in non-liberal democracy), or what I originally called "authoritarian liberal democracy" (as summarized in *Table 4.19*), before it is to be superceded by "post-democracy."

In regard to economics, I also offered another vision, this time to understand the future forms of economic systems, that is, the different forms of "post-capitalism" to supersede capitalism unto the post-human age.

My vision of post-capitalism was referred to (in *FCD*, *BDPD* and *BCPC*) as "the theory of post-capitalism" (as summarized in *Table 4.26*, *Table 4.27*, *Table 4.28*, *Table 4.29*, and *Table 4.30*). And *Table 4.31* clarifies the differences among capitalism, non-capitalism, and post-capitalism, whereas *Table 4.32* shows multiple causes of the emergence of post-capitalism.

This vision on "post-capitalism" constitutes the third theoretical application of existential dialectics, this time, in that there is no wealth without poverty (or the wealth/poverty dialectics).

More importantly, these existential constraints hold true for the age of pre-modernity, modernity, post-modernity, and, in the future, what I originally suggested in *FHC* as "after-postmodernity"—as shown in *Table 1.32*, *Table 1.33*, *Table 1.34*, and *Table 1.35*.

This is all the more so, in the "post-human" age at some distant point of "after-postmodernity," long after human extinction, to be eventually superseded by post-humans of various forms. Some good candidates of the post-human lifeforms include, for instance, "thinking robots," "thinking machines," "cyborgs," "genetically altered superior beings," "floating consciousness," and "hyper-spatial consciousness." This post-human vision of mine was first originally worked out in *FHC* and further elaborated in both *FCD* and *FPHC*, for instance.

Lest any confusion occurs, I need to emphasize two clarifications here about the term "post-human" as a neologism in my works.

Firstly, the word "post-human" here should *not* be confused with another term which looks similar but has a totally different meaning in the literature of postmodernism, namely, "post-humanism"—which constitutes a critique of "humanism" as tradionally understood (especially, though not exclusively, in relation to the idea of progress in science and reason in the Enlightenment project). (WK 2008)

My works reject the project of "postmodernism" and propose the future world of what I orignally called "after-postmodernity" in *FHC* and *FCD*, for instance.

And secondly, the word "post-human" here should also *not* be confused with a similar term which is used to champion the ideology of technology for the future co-existence between humans and cyborgs in "trans-humanism." (WK 2008a)

Instead, my term "post-human" in relation to "posthuman-ism" also rejects "transhumanism" (especially, though not exclusively, in relation to the promises of technology) and refers to something else altogether, that is, the future extinction of humans and its post-human successors in deep space and beyond unto multiverses.

With these two clarifications in mind (as summarized in *Table 1.31*)—in the end, a most fundamental question about intelligent life now has an answer, in that, if asked, "What is the future of human civilization?"—my answer in *FCD* (89) is thus: "As addressed in Ch.7 of *FHC*, a later epoch of the age of after-postmodernity (that is, at some point further away from after-postmodernity) will begin, as what I called the 'post-human' history (with the term 'post-human' originally used in my doctoral dissertation at M.I.T., which was finished in November 1995, under the title *After Postmodernity*, still available at M.I.T. library, and was later revised and published as *FHC*). The post-human history will be such that humans are nothing in the end, other than what culture, society, and nature (with some luck) have shaped them into, to be eventually superseded by post-humans (e.g., cyborgs, thinking machines, genetically altered superior beings, and others), if humans are not destroyed long before then."

A conclusion which is shocking indeed for many contemporaries is none other than that "[t]he post-human history will therefore mark the end of human history as we know it and, for that matter, the end of human dominance and, practically speaking, the end of humans as well. The entire history of human civilization, from its beginning to the end, can be summarized by four words, linked by three arrows (as already discussed in *FHC*)":

Pre-Modernity → Modernity →
Post-Modernity → After-Postmodernity

In *BDPD*, this thesis of mine was specifically called "the theory of the evolution from pre-modernity to after-postmodernity," at the historical level.

Therefore, "[t]he end of humanity in the coming human extinction is the beginning of post-humanity. To say an untimely farewell to humanity is to foretell the future welcome of post-humanity." (P. Baofu 2002: 89) This thesis of mine was known in *BDPD* as "the theory of post-humanity," at the systemic level.

In Ch.9 of *FCD* (367-8), I further showed "that civilizational history will continue into the following cyclical progression of expansion, before it is to be superseded (solely as a high probability, since humans might be destroyed sooner either by themselves or in a gigantic natural calamity) by posthumans at some distant point in after-postmodernity (which I already discussed in *FHC*)" unto multiverses (different constellations of universes):

Local → Regional → Global →Solar →
Galactic → Clustery…→ Multiversal

In *BDPD*, this thesis of mine was referred to as "the theory of the cyclical progression of system integration and fragmentation," at the systemic level—in close relation to another thesis analyzed in Ch.9 of *FCD*, which was called, in the absence of better words, "the multifaceted theory of war and peace" in *FPHK*.

In *BWT*, a different version of this thesis is "the theory of cyclical progression of empire-building," at multiple levels (e.g., institutional, structural, systemic, and others), in providing a better way to understand the logic of empire-building on earth and beyond—with the rise of what I originally envisioned as "hyper-empires" (like "the Chinese Union" and "the Indian Union"), "meso-empires" (like "the European Union" and "the North American Union") and "micro-empires" (like "the Middle Eastern Union" and "the Latin American Union") in the coming post-post-Cold War era as a visionary illustration, as summarized in *Table 4.18*.

In *BNN*, I further worked out "the theory of contrastive advantages" (which was originally proposed in *FCD* and summarized in *Table 1.24*), to show the nature-nurture interactions of multiple levels in action (e.g., the biological, the psychological, the structural, the systemic, the cultural, and others) for humans and post-humans on earth and beyond, to the extent that different groups, be they on the basis of race, gender, ethnicity, class, age, or else, are not equal, on average, and have never been, nor will they be, even if there can be exceptions.

In fact, even greater transformations are yet to come in the post-human age. For instance, even the existence of human consciousness will be superseded one day too, with "floating consciousness" (as first proposed in *FCD*) and "hyper-spatial consciousness" (as first proposed in *FPHC*) as a climax of evolution in consciousness, after the future extinction of human consciousness (as shown below):

Primordial consciousness → Human consciousness →
Post-human consciousness (with floating consciousness and hyper-spatial consciousness as a climax in the evolution of consciousness)

In *BDPD*, these latest theses of mine were called "the theory of floating consciousness" and "the theory of hyper-spatial consciousness," both at the cosmological and psychological levels (as summarized in *Table 1.26* and *Table 1.25*).

In *FPHC*, the nature of consciousness was further detailedly analyzed, especially in relation to the conceptual dimensions of consciousness (as shown in *Table 1.12*), the theoretical levels of consciousness (as shown in *Table 1.13*), the thematic issues of consciousness (as shown in *Table 1.14*), the different dimensions of human existence (viz., having-ness, belonging-ness, and being-ness) in relation to consciousness (as shown in *Table 1.15*, *Table 1.16*, *Table 1.17*, and *Table 1.18*).

In *BCPC*, the nature of consciousness was explored further in relation to cognition, emotion, and behavior, which can be in consciousness and other mental states (e.g., subconsciousness and unconsciousness), in the context of existential dialectics, as shown in *Table 1.19*, *Table 1.20*, *Table 1.21*, *Table 1.22*, and *Table 1.23*.

But this is not the end of the matter yet.

In *FPHU*, even the unconsious mind will undergo tremendous change, to the rise of what I originally called "unfolding unconsciouess" in relation to anomalous experience, as a climax of evolution in unconsciousness, in conjunction with the conscious mind (both human now and post-human later), especially at the psychological level, as shown below:

Primordial unconsciousness → Human unconsciousness →
Post-human unconsciousness (with unfolding unconsciousness as a climax in the evolution of unconsciousness)

In *FPHU*, this thesis of mine is called "the theory of unfolding unconsciousness" in a small sense, or "the unfolding theory of anomalous experience," in a larger one, as shown in *Table 1.27* and *Table 1.28*.

This thesis is also related to another thesis on religion which was already analyzed in Ch.3 of *FHC* and Ch.9 of *FCD* but was referred to, in the absence of better words, as "the comparative theory of religion" in *FPHK*.

In *FAE*, I also showed how and why this technological direction will fundamentally transform the nature of aesthetic experience.

In *FPHE*, I further analyzed the nature of technology itself, in the context of engineering.

In *FPHMM*, I addressed this technological future from the vantage point of communication.

In *FPHCT*, I explored the role of creative thinking, in relation to technology and other factors, together with my original proposal of important creative techniques and traits (as shown in *Table 1.29*) and its double nature of desirability and undesirability (as shown in *Table 1.30*).

In *FPHG*, I link this technological concern with mathematics and other disciplines, with implications for the understanding of space-time in the context of infinity, symmetry, and dimensionality for future lifeforms that our world has never known (as shown in *Table 1.6*, *Table 1.8*, *Table 1.9*, *Table 1.10*, and *Table 1.11*).

*The Further Use of Existential Dialectics*

Even more interestingly, in *BDPD* and later in *BCIV*, I further revealed a theoretical refinement of existential dialectics, this time, in arguing that there is no civilization without barbarity, with human civilization to be eventually superseded by what I originally analyzed as "post-human post-civilization" (which should not be confused with "post-human civilization"), in the context of the freedom/unfreedom and equality/inequality dialectics.

In the final analysis, civilization cannot live without barbarity and has to learn to co-exist with it in ever new ways. Preposteorus as this may seem to many contemporaries—it is no more imperative to preserve civilization than necessary to destroy barbarity, and the ideal of civilization is essentially bankrupt, to be eventually superseded by "post-civilization" (as summarized in *Tables 4.33-4.41*).

Yet, this requires some understanding of my analysis of the trinity of modernity and other ages. For instance, in both *FCD* and *FPHC*, I worked out the structure of "post-human civilization" in terms of the trinity of after-postmodernity (i.e., "free-spirited after-postmodernity," "post-capitalist after-postmodernity," and "hegemonic after-postmodernity").

Both conceptually and theoretically, the trinity of after-postmodernity is a sequential extension to the trinity of modernity (i.e., "free-spirited modernity," "capitalist modernity,"and "hegemonic modernity") and the trinity of postmodernity (i.e., "free-spirited postmodernity," "capitalist postmodernity," and "hegemonic postmodernity") as first proposed in *FHC*.

And the trinity of pre-modernity (i.e., "pre-free-spirited pre-modernity," "pre-capitalist pre-modernity" and "hegemonic pre-modernity") was later conceived in *BCIV* to complete the historical set from pre-modernity to after-postmodernity.

In *BDPD*, this thesis about the trinity of pre-modernity, modernity, postmodernity, and after-postmodernity was collectively known as "the theory of the trinity of modernity to its after-postmodern counterpart," at the cultural level (as summarized in *Table 1.32*, *Table 1.33*, *Table 1.34*, and *Table 1.35*).

At the structural level, all these trinities are subject to the existential constraints (e.g., the freedom/unfreedom and equality/inequality dialectics in the context of "the cyclical progression of hegemony"), be the historical epoch in pre-modernity, modernity, postmodernity, or after-postmodernity in future times. Each of the historical epochs has its ever new ways of coming to terms with the ever new (different) mixtures of freedom/unfreedom and equality/inequality.

This is importantly so, not because, as is falsely assumed in conventional wisdom, one certain way is superior (or better) than another in terms of achieving more freedom and less unfreedom, or more equality with less inequality.

On the contrary, indeed, in each of the historical epochs, each increase of unfreedom greets each freedom achieved, just as each increase of inequality welcomes each equality achieved, albeit in ever new (different) ways. In *BDPD*, this thesis of mine was labeled as "the theory of the cyclical progression of hegemony," at the structural level, though it was first analyzed in *FCD*.

In *BDPD*, more theoretical applications of existential dialectics were further examined, in relation to five main features, in the context of the duality of oppression, namely, (a) that each freedom/equality achieved is also each unfreedom/inequality created, (b) that the subsequent oppressiveness is dualistic, both by the Same against the Others and itself and by the Others against the Same and themselves, (c) that both oppression and self-oppression can be achieved by way of downgrading differences (between the Same and the Others) and of accentuating them, (d) that the relationships are relatively asymmetric among them but relatively symmetric within them, even when the Same can be relatively asymmetric towards itself in self-oppression, and the Others can be likewise towards themselves, and (e) that symmetry and asymmetry change over time, with ever new players, new causes, and new forms, be the locality here on Earth or in deep space unto multiverses—as summarized in *Table.4.12*, *Table 4.13*, *Table 4.14*, and *Table 4.15*.

The same logic also holds both in relation to wealth and poverty (as addressed in *BCPC* and summarized in *Table 5.16* on the wealth/poverty dialectics) and in relation to civilization and barbarity (as addressed in *BCIV* and summarized in *Table 5.17* on the civilization/barbarity dialectics).

In *BDPD*, this thesis on existential dialectics was labeled as "the theory of existential dialectics."

*Direct and Indirect Applications of Existential Dialectics*

A different way to appreciate the usefulness of existential dialectics is by way of the analysis of its (a) direct and (b) indirect applications.

(a) In direct applications, on the one hand, the logic of existential dialectics can shed some theoretical insights on diverse phenomena in the world, and good instances are the usage of the principles of existential dialectics for the theoretical insights on the freedom/unfreedom dialectics, the equality/inequality dialectics, and the wealth/poverty dialectics (as introduced above).

For instance, my books like *FPHST*, *BNN*, *FAE*, *FC*, *FIA*, *FPHU*, *FPHK*, *FPHML*, *FPHE*, *FPHMM*, *FPHCT*, *FPHG*, and *FPHUP* use the principles to reveal some theoretical insights on the perspectives of space and time (as in *FPHST*), of nature and nurture (as in *BNN*), of beauty and ugliness (as in *FAE*), of order and chaos (as in *FC*), of taxonomy and network (as in *FIA*), of the unconscious mind (as in *FPHU*), of the nature of knowledge (as in *FPHK*), of mathematical logic (as in *FPHML*), of engineering (as in *FPHE*), of mass media (as in *FPHMM*), of creative thinking (as in *FPHCT*), of geometry (as in *FPHG*), and of urban planning (as in *FPHUP*).

(b) In indirect applications, on the other hand, the theoretical insights can further be used to reveal other phenomena directly from them (viz., the theoretical insights) and therefore indirectly from the principles themselves.

A good illustration is of course the use of the theoretical insights on the freedom/unfreedom and equality/inequality dialectics for the understanding of the civilization/barbarity dialectics.

This distinction between direct and indirect applications may sound a bit academic, since even in indirect applications, the phenomena under study can still be directly related back to the principles themselves.

In the previous example, as an illustration, the civilization/barbarity dialectics can be directly related to the principles of existential dialectics without the intermediate role of the freedom/unfreedom and equality/inequality dialectics.

*Multiple Levels of Application*

There is another issue to be clarified, however. In other words, the theoretical insights can be applied to multiple levels of analysis—even though, in a given example, it may refer to one level or a few only.

For instance, in the example concerning the freedom/unfreedom dialectics, it can be used at the structural level (e.g., in relation to the theory of the cyclical progression of hegemony), but it can also be exploited for other levels (e.g., the theory of post-capitalism at the institutional level).

All these levels of application should not be misleadingly construed, as I stress this before, as a one-way street (that is, to use the ontological principles for theoretical application at multiple levels of specific analysis) but a two-way one, in which theoretical insights in praxis, when studied in more specific contexts, can in turn refine the nature of existential dialectics (for example, with the addition of new principles).

## Sophisticated Methodological Holism

The summary of my original meta-theory on methodology is provided below, again, to be repeated *often verbatim* from my previous works—as this is something I regularly do in each new book to introduce my past works, which are in conversation with this current one.

This means that, for those readers who had read some of my previous books, the summary below is really nothing new (*almost verbatim*, with only some minor revisions) and can be skipped.

That clarified—this original contribution of mine to the study of methodology takes the form of an approach known as "the theory of methodological holism" or "methodological holism" in short, as already worked out in *FPHC* (2004).

I have made good use of this methodological approach of mine for all of my previous works.

Yet, it is imperative to stress at the outset, as I have so emphasized in all my previous books, that my approach of "methodological holism" does not oppose or exclude "methodological individualism" (as some reader may be tempted to assume, as is conventionally understood) but actually includes it.

For this reason (and others too, as summarized in *Table 4.1*), my version of methodological holism is *sophisticated*—not *vulgar* as sometimes used by inapt scholars employing the same term.

*The Ontological Constraints*

Sophisticated methodological holism is subject to some ontological constraints, and five good examples include "the partiality-totality principle," "the predictability-unpredictability principle," "the explicability-inexplicability principle," "the absoluteness-relativeness principle," and "the formalness-informalness principle" in existential dialectics.

Of course, there can be more than five ontological principles in relation to method, but the five examples here are sufficient to illustrate the point in question.

With this caveat in mind—the first two were already analyzed in previous books of mine like *FC* and *FAE*, whereas the third one was addressed in *FPHU*, and the last two were accounted for in *FPHK* and *FPHML*, respectively.

For now, it suffices to show that sophisticted methodological holism, because of these ontological constraints on methodology, targets two major sins of methodology, namely, what I call, in the absence of better words, (a) *reductionism* and (b) *reverse-reductionism*—both of which come in all shapes and sizes, to be summarized below (and also shown in *Table 4.2*).

*Against the Varieties of Reductionism*

There are four versions of reductionism to be summarized here, which sophisticated methodological holism rejects, namely, (a1) conceptual, (a2) theoretical, (a3) methodological, and (a4) ontological forms of reductionism.

(a1) In conceptual reductionism, a good case in point concerns myriad dualities like mind vs. body, self vs. world, democracy vs non-democracy, and the like (as already addressed in *FHC*, *FPHC*, and *BDPD*, for instance).

(a2) In theoretical reductionism, an illuminating case study is best exemplified by what I originally called "the foundation fallacy" in *FPHST*, in any attempt to naively understand space-time from the physical perspective as the foundation and, consequently, to dangerously dismiss (or belittle) other perspectives.

In *FAE*, I elaborated further these versions of reductionism in the literature on aesthetics (e.g., form vs. content, representation vs. expression, critics vs. artists, and externalism vs. internalism).

In *FIA*, I revealed other forms of reductionism in the literature on information architecture (e.g., the constructivist argument).

In *FPHU*, I showed the persistent legacy of reductionism, this time, in the literature on anomalous experience (e.g., the obsession with physics, chemistry, and biology for explaining anomalous experience).

In *FPHE*, I examined another case of reductionism in action, in the context of engineering (e.g., technical constraints vs. normative constraints).

In *FPHMM*, I elaborated one more version of reductionism, in relation to the three domains of communication (e.g., the competing views on sending, connecting, and receiving).

In *FPHCT*, I explored another version of reductionism, in relation to invention and innovation (e.g., the bio-psychological argument vs. the socio-cultural arguments).

In *FPHG*, I identified another version of reductionism, in relation to infinity, symmetry, and dimensionality (e.g., the Euclidean argument vs. the non-Euclidean arguments).

And in *FPHUP*, I analyzed another version of reductionism, in relation to density and void (e.g., the engineering argument and the ecology argument).

(a3) In methodological reductionism, a good illustration can be the debate between different versions of qualitative and quantitative methods (as already analyzed in *FC* and also *FHC*).

In *FPHML*, I examined similar reductionism, this time, in the literature on mathematical logic (e.g., the obsession with consistency, soundness, and completeness).

(a4) In ontological reductionism, an excellent instance involves another debate, this time between emergentism and reductionism in complexity theory (as addressed in *FC*) and also in psychology (as elaborated in *FPHC*, in the context of Being and Belonging).

But to target against the varieties of reductionism constitutes only a side of the same coin, and the other side concerns the varieties of reverse-reductionism.

### *Against the Varieties of Reverse-Reductionism*

The other side of the same coin is the reverse version of reductionism, which is what I want to call, in the absence of better words, *reverse-reductionism*.

My sophisticated version of methodological holism targets against the varieties of reverse-reductionism (just as it also rejects the ones of reductionism).

Perhaps nothing expresses better the popularity of reverse-reductionim than the "anything-goes" mentality in postmodernism of our time, as shown in the following four versions.

(b1) In conceptual reverse-reductionism, any concept of "art" (e.g., fine art, applied art, outsider art, junk art) is welcome in postmodernism (as already analyzed in Ch.4 of *FHC*—and also in *FAE*).

(b2) In theoretical reverse-reductionism, a variety of art and literary theories co-exist. Take the case of literature, as there are now Literary Structuralism, Marxist Literary Criticism, New Criticism, Phenomenology, Hermeneutics, Language-Game Literary Criticism, Feminist Literary Criticism, Reception Theory, Reader Response Criticism, Poststructuralism, Semiotics, Pyschoanalytic Literary Criticism, just to cite some well-known ones, with no one being said to be better than any others (as detailedly analyzed in Ch.4 of *FHC*). (S. Raman 1997) In *BNN*, I even introduced "the compromise fallacy" as another good illustration of theoretical reverse-reductionism, in misleadingly treating both genetic and environmental approaches as equally valid.

(b3) In methodological reverse-reductionism, multiple methodologies are deemed as acceptable in postmodernism (e.g., doing art without praxis, doing art with praxis, and doing art by sublation), as analyzed in Ch.4 of *FHC*.

(b4) In ontological reverse-reductionism, no privileged ontology is allowed, and the door is open practically for anything in postmodernism (e.g., the equal status of the ontology of Being vs. that of Becoming, as already addressed in Ch.4 of *FHC*—and also in *FPHC*).

In *FAE*, I also introduced another version of reverse-reductionism, that is, "the pluralist fallacy," in the context of understanding aesthetic experience, for instance—although this fallacy has been committed not exclusively in relation to the ontological level (but also at the conceptual, theoretical, and methodological ones).

These dual dangers against reverse-reductionism (in this sub-section) and reductionism (in the previous sub-section) are something that sophisticated methodological holism rejects.

This point is important enough, since many scholars often favor one against the other, but it is vital to target both.

*The Holistic Organization of an Inquiry*

With these dual dangers against reductionism and reverse-reductionism in mind—sophisticated methodological holism suggests that an inquiry of any given phenomenon is more complete, if treated in the context of a comprehensive analysis at all relevant levels, which challengingly encompass all the fields of human knowledge, ranging from the natural sciences and formal sciences through the social sciences to the humanities.

This section is something that I had already stressed in all my previous books and repeat (*often verbatim*) hereafter. But, for those readers who had read my previous books, this serves as a reminder.

With this reminder in mind—there are multiple ways to engage in a holistic inquiry with all relevent levels of analysis. Over the years, I have proposed different ways to fulfill this holistic methodological requirement, as repeated in my previous books.

Hereafter is a summary of four major ways, namely, (i) by discipline, (ii) by domain, (iii) by subject, and (iv) by meta-analysis—all of which fulfill the holistic requirement of sophisticated methodological holism on a given issue.

(i) In a holistic organization by discipline, a good illustration is the multiple levels of analysis, namely, (i1) the micro-physical, (i2) the chemical, (i3) the biological, (i4) the psychological, (i5) the organizational, (i6) the institutional, (i7) the structural, (i8) the systemic, (i9) the cultural, (i10) the macro-physical (cosmological), and (i11) other relevant levels, which are either a combination of all these levels or the practical applications with a combination of them.

This holistic organization by discipline is used in *FCD* and *BCPC*, for instance.

(ii) In a holistic organization by domain, some good candidates include the classification of different perspectives of analysis in relation to (ii1) nature, (ii2) the mind, (ii3) society, and (ii4) culture, as already worked out in *BCIV*, *FPHST*, *FC*, *FAE*, *FPHU*, *FPHK*, and *FPHE*.

In this re-classification, culture in (ii4) is the same as culture in (i9) and can be further divided into different sub-categories, with good examples like (i9a) epistemology, (i9b) aesthetics, (i9c) ethics, and (i9d) religion.

Society in (ii3) includes the organizational in (i5), the institutional in (i6), the structural in (i7), and the systemic in (i8). The mind (ii3) has more to do with the chemical in (i2), the biological in (i3), and the psychological in (i4), although it can overlap with (i1), (ii3) and (ii4), for instance. And nature in (ii1) refers to the micro-physical in (i1) and the macro-physical (cosmological) in (i10), although it can also overlap with (i2), (i3) and (i4).

(iii) In a holistic organization by subject, some excellent examples concern the analysis of consciousness in *FPHC* (i.e., on Having, Belonging, and Being); of civilization in *FHC* (i.e., the True, the Holy, the Everyday, the Technological, the Beautiful/Sublime, the Good, and the Just); of international politics in *BWT* (i.e., Hyper-Empires, Meso-Empires, and Micro-Empires); of nature and nurture in *BNN* (i.e., genes and memes); of taxonomy and network in *FIA*; of methodology and ontology in *FPHK*; of consistency, soundness, and completeness in *FPHML*; of sending, connecting, and receiving in *FPHMM*; of invention and innovation in *FPHCT*; of infinity, symmetry, and dimensionality in *FPHG*; and of density and void in *FPHUP*—by incorporating all the relevent levels of analysis as cited above in each.

(iv) In a holistic organization by meta-analysis, two good cases in point involve the works on post-capitalism in *BCPC* and on authoritarian liberal democracy in *ALD*, with the classification of analysis in theory and meta-theory.

The essential point to remember here, as I thus stress in all my previous books, is that the multiple levels of analysis can be reorganized in many different ways, insofar as none of the levels (if relevant to an inquiry in question) is ignored or dismissed, to avoid the danger of reductionism (and for that matter, the one of reverse-reductionism, in the opposite direction).

### *Three Clarifications*

Three clarifications are needed here, however, to avoid any hasty misunderstanding.

Firstly, the entities in each classification are not mutually exclusive. For instance, in the holisitc organization by domain, nature in (ii1) can also be linked to the chemical in (i2), the biological in (i3), and the systemic in (i8). And the mind (ii3) can alternatively be related to the micro-physical in (i1) and the chemical in (i2), for example.

Secondly, there is also the important factor of luck (or uncertainty in conventional usage), but it is already implied or allowed in each of the organizations of an inquiry—especially in relation to the predictability-unpredictability principle in existential dialectics.

And lastly, the comparison in a classification is not absolute, but relative, as there are often some shades of gray, not exactly black or white (figuratively speaking).

Some further clarifications and qualifications of sophisticated methodological holism are shown in *Table 4.1*.

*Some Distinctive Usefulness*

In the end, sophisticated methodological holism—when applied, especially though not exclusively, as illustrated in my numerous works—can enrich the understanding of reality in some distinctive ways. Here are three examples (as revealed in each of my books).

Firstly, it provides a comprehensive analysis of a subject matter, from which much can be learned about reality, in relation to the perspectives of the mind, nature, society, and culture.

Secondly, it suggests some visions of the future in relation to the subject matter in question.

And thirdly, it proposes some insights on meta-theory (e.g., methodology and ontology) in general—with the clear understanding, however, of the dilemma of specific vs. general ontology (as shown in *Table 4.9* on the syntax of existential dialectics in the context of the dilemma of ontology).

For this very reason, all of these ways are important, without reducing one into the analysis of another." (*FPHCT*)

# Chapter Outline

With this summary of my two original meta-theories (on methodology and ontology, *almost verbatim* from my previous works) in mind—this book is organized in four main parts, namely, (a) Part I on the introduction to this book, (b) Part II on density, (c) Part III on void, and (d) Part IV on the conclusion of the book.

The book is thus organized in four chapters.

The introductory chapter here, that is, Chapter One titled "Introduction—The Legacy of Urban Planning," starts with the issue concerning the obsession with sustainability in contemporary urban planning, the metamorphic history of urban planning, and the relationships among density, void, and sustainability, together with the theoretical debate in the literature, my contingent theory of urban planning, the relationship between theory and meta-theory, existential dialectics, sophisticated methodological holism, and some further clarifications.

Chapter Two titled "Density and Its Crowdedness" examines the nature of density in urban planning from the perspectives of culture, society, nature, and the mind, with a critical evaluation of each.

Chapter Three titled "Void and Its Solitude" then analyses the nature of void in urban planning from the perspectives of culture, society, nature, and the mind, with a constructive analysis of each.

The last chapter titled "Conclusion—The Future of Urban Planning" summarizes the analysis in the book with the seven theses, namely, (a) the first thesis: the denseness-emptiness principle, (b) the second thesis: the absoluteness-relativeness principle, (c) the third thesis: the simpleness-complicatedness principle, (d) the fourth thesis: the evolution-transformation principle, (e) the fifth thesis: the symmetry-asymmetry principle, (f) the sixth thesis: the regression-progression principle, and (g) the seventh thesis: the post-human planning—to be elaborated in the rest of the book, with a summary in the concluding chapter.

If true, this seminal theory shows that the concern with sustainability, with the analysis of density and void in urban planning as a case study, has been much exaggerated and distorted, such that its dark sides have been unwarrantedly ignored or downgraded.

Therefore, this new theory will fundamentally change the way that we think about the issue of sustainability by transcending it, with enormous implications not only for understanding the future of urban planning, in a small sense—but also for predicting the relevance of sustainability in relation to the entire domain of human knowledge for the human future and what I originally called its "post-human" fate, in a broad sense.

## Some Clarifications

But some clarifications are needed, before a formal analysis of the nature of post-human urban planning is to begin in Chapter Two.

These clarifications, lest any potential misunderstanding by the readers occurs, refer to (a) the conversation with my previous books, (b) the illustrative selection of case studies and examples, (c) detailed analysis versus overall analysis, and (d) the use of neologisms.

These clarifications are something that I also regularly repeat in my previous books too, so these points are repeated hereafter, *often verbatim*—albeit in a different context this time.

### In Conversation with My Previous Books

Firstly, my contingent theory of urban planning is constructed on the theoretical foundation of my previous books (i.e., *FHC*, *FCD*, *FPHC*, *BDPD*, *BCPC*, *BCIV*, *FPHST*, *BNN*, *BWT*, *FC*, *FAE*, *ALD*, *FIA*, *FPHU*,

*FPHK*, *FPHML*, *FPHE*, *FPHMM*, *FPHCT*, and *FPHUP*—as explicated in "The List of Abbreviations," on page xix).

Although a summary of my previous books is provided whenever necessary, as shown in the two sections above (*often verbatim*) and in different tables at the end of some chapters, I still expect the readers to read my previous books directly for more details.

## Case Studies and Examples

Secondly, the case studies and examples in the chapters on density and void are not exhaustive but solely illustrative. Nor are they necessarily mutually exclusive, as they can be reclassified in a different way. And exceptions are allowed.

One is tempted to assume, however, that, since the case studies and their examples are not exhaustive, more case studies and examples are needed. But the problem here is that just adding more of them does not necessarily change much of the conclusions to be drawn and may even lead to redundancy.

After all, the case studies and examples used in this book are chosen with care and deemed sufficient for the purpose at hand, even though they are solely illustrative.

## Detailed Analysis versus Overall Analysis

Thirdly, there is a distinction between detailed analysis and overall analysis, in that the former is more for the specialized viewpoint of a specific discipline, whereas the latter is more for the broad horizon of an interdisciplinary approach—relatively speaking, of course.

Or, by analogy, there is a distinction between depicting a "tree" (with tiny details) in a "forest" and viewing the entire "forest" itself (with the broader horizon).

This book, like all previous ones of mine, tries to strike a balance between the two forms of analysis, although this does not imply that the two forms should be exactly equally offered, in an exact 50-50 balance.

Instead, the book, more often than not, does not allow the tiny details to obscure the larger picture in a given case study, because the entire book is inter-discipliary in nature and tries to look at the entire "forest," not being bogged down with the detailed analysis of an individual "tree" in the forest, by analogy.

In this sense, although detailed analysis can be provided whenever needed for a given case study, the important point to remember is the overall horizon with its broadness of scope.

## The Use of Neologisms

And fourthly, I use different neologisms in my books, mostly to introduce my original concepts and theories, and good instances include those here or elsewhere in my previous books (e.g., "the contingent theory of urban planning," "the denseness-emptiness principle," "post-democracy," "hyper-spatial consciousness," "post-capitalism," and whatnot). Needless to say, they are used here solely for our current intellectual convenience, as they will be renamed differently in different ways in future history.

As I thus wrote in *FCD* (508-9), "all these terms 'post-capitalism,' 'post-democracy'...and other ones as introduced in...[the] project (e.g.,...'posthuman elitists,' and 'posthuman counter-elitists,' just to cite a few of them) are more for our current intellectual convenience than to the liking of future humans and post-humans, who will surely invent more tasteful neologisms to call their own eras, entities, and everything else, for that matter. But the didactic point here is to use the terms to foretell what the future might be like, not that its eras and entities must be called so exactly and permanently. After all, William Shakespeare (1995: Act II, Scene II, Line 47) well said long ago: 'What is in a name? That which we call a rose by any other name would smell as sweet.'"

As I had also stressed time and again before, each of the neologisms can be re-written as a different "X," only to be re-named differently by the powers that be in different eras of future history.

With these clarifications for thought in mind—I now proceed to Chapter Two on density and its crowdedness.

**Table 1.1. Density, Void, and Sustainability (Part I)**

- **Density**
  —The term "density," in its conventional usage, refers to "the quantity per unit volume, unit area, or unit length," "the distribution of a quantity…per unit usually of space (as length, area, or volume)," or "the average number of individuals or units per space unit." (MWD 2008)
  —In the context of urban planning, "density" is used here to refer to the distribution of entities (e.g., inhabitants, buildings, roads, parks, etc.) in a given city (or area, in general) over time.
  —The issue of "density," when contextualized in relation to sustainability in urban planning, concerns different phenomena (e.g., "traffic density, road rage, and human emotion"; "urban density and mental health"; "ecological complexity, and the urban balance with nature"; "urban pollution and institutional failures"; "urban renewal and structural injustice"; "sustainable technology and system complexity"; "urban planning and the religious search for order and purity"; "aesthetics and soundscape planning"; and the like).

*(continued on next page)*

**Table 1.1. Density, Void, and Sustainability (Part II)**

- **Void**
  - —By contrast, the term "void," in its conventional usage, refers to something which is "not occupied," "not inhabited" (or "deserted"), or simply "empty," relatively speaking, of course. (MWD 2008a)
  - —The qualification of "relatively speaking" above is used with good reason, since, in physics, it is well known that "empty" space is not really empty all the way, because it can be full of energy (e.g., random quantum fluctuations at the sub-atomic level, and, for that matter, dark energy in the universe) and matter (e.g., different versions of sub-atomic particles, and, for that matter, dark matter in the universe). (F. Wilczek 2008)
  - —Therefore, "it is practically impossible to construct a region of space that contains no matter or fields, since gravity cannot be blocked and all objects at a non-zero temperature radiate electromagnetically." (WK 2009j)
  - —With this qualication in mind—the lack of density is "void," even if used here merely as an "approximation," depending on the degree of the lack of density.
  - —In the context of urban planning, "void" here therefore means the relative absence of the distriubtion of entities (e.g., inhabitants, buildings, roads, parks, etc.) in a city (or area, in general) over time, such that it is not occupied, not inhabited, or empty—again, relatively speaking.
  - —The issue of "void," when contextualized in relation to sustainability in urban planning, then refers to different phenomena (e.g., "indifference and urban solitude"; "urban homelessness and mental Health" ; "deforestation and urban destructiveness": "urban open space and its controversy"; "urban decay and its cycle"; "urbicide and urban warfare"; "individualism and urban emptiness"; "the monastic emptiness, and urban suffering"; and so on).

*(continued on next page)*

**Table 1.1. Density, Void, and Sustainability (Part III)**

- **Sustainability**
  —The term "sustainability," in its conventional usage, refers to "a method of harvesting or using a resource so that the resource is not depleted or permanently damaged" over time. (WMD 2008b)
  —In the technical jargons of economics, for instance, sustainability strives for "optimization" (for the first best solution, or something close to it), not just "maximization" (for the second best solution, or something close to it, without dealing with, say, negative externalities like environmental pollution).
  —In the current context of urban planning (and other related fields), "sustainability" has a technical meaning, in relation to four main features, namely, (a) temporal persistence, (b) ecological interdependence, (c) moral intergenerational equity, and (d) policy implications. (M. Dempster 1998)
  —In fact, other scholars go a step further and argue that sustainability should also include other serious issues like "spiritual enligthenment" (to reduce urban existential suffering) and "social justice" (for the poor, minorities, and other unfortunate ones in society)—as will be clear in the next section and the rest of the book.
  —Unfortunately, not every scholar in urban planning agrees with the policy implications of the main features of sustainability.

*Sources*: A summary of *Sec.1.3* in *FPHUP*

**Table 1.2. The Theoretical Debate on Urban Planning (Part I)**

- **The Engineering Argument**
  - —Firstly, a major theoretical approach can be called, in the absence of better words, as the *engineering* argument, which advocates a top-down approach towards urban planning, especially though not exclusively with the use of technology.
  - —The engineering argument refers to the "systems engineering" tradition which can be traced back to the "planning beginning with Saint Simon and August Comte in the early 1800s. There is a pervasive, 'systems thinking' influence throughout the history of planning that is based on the engineering control paradigm." (M. Dempster 1998)
  - —In the later part of the $20^{th}$ century, "addition of systems analysis to the rational model…has created the sense of a more exact, scientific approach….This particular approach has been integral to planning, especially within institutional frameworks, though to a substantially lesser degree in radical forms of planning." (M. Dempster 1998; J. Wolfe 1989)
  - —A good illustration concerns the "strategic urban planning" (SUP) which suggests "clarifying which city model is desired and working towards that goal, coordinating public and private efforts, channelling energy, adapting to new circumstances and improving the living conditions of the citizens affected." (WK 2008j)
  - —For instance, the city of San Francisco in California "carried out…[the] process [of SUP] between 1982 and 1984. The main motivation behind starting strategic urban planning processes was the attempt to adequately react to problematic situations (mainly economic crisis or standstill)." (WK 2008j)
  - —For the critics, however, the top-down approach in the engineering argument is elitist, in that it does not pay sufficient attention to "civic participation" and "political activism" for the less well offs in the city. (WK 2008j)

*(continued on next page)*

**Table 1.2. The Theoretical Debate on Urban Planning (Part II)**

- **The Engineering Argument (*cont'd*)**
  —Another serious problem with the "engineering control paradigm" for urban planning is the inherent "difficulties of societal guidance," as John Friedmann (1987 & 1996) thus said: "Because it is invariably integrated into the state apparatus, planning for societal guidance is incapable of coping with the crisis of industrial capitalism." (M. Dempster 1998)

- **The Ecology Argument**
  —Secondly, another theoretical approach can be referred to, in the absence of better terms, as the *ecology* argument, in that, unlike the engineering argument, it pays more focus on the close linkages between the ecological system and the human environment in urban planning.
  —In fact, the ecology argument rejects the greedy presumption of "profit maximization" so much championed in lasses-faire capitalism (as famously shown in the magic of the "invisible hand" by Adam Smith and its successors in the neo-classical economic tradition), which often ignores or does not pay serious attention to the problem of ecological destruction in the classical developmental process, which is thus criticized as un-sustainable. As M. Beth Dempster strongly argued, "the neoclassical economic interpretation…typically fails to recognize our dependence on environmental inputs….This is particularly important when it is noted that societies that perceive their resources as infinite are not likely to reduce their impacts….This is especially problematic when it is recognized that a society can misinterpret the state of their resources." (M. Dempster 1998; M. Gadgil 1991; E. James 1991)
  —This problem of unsustainability is all the more serious, when it has now reached the global scale of greenhouse effects, to the point that "we are talking about our planetary survival system." (M. Dempster 1998; K. McNamee 1996)

*(continued on next page)*

**Table 1.2. The Theoretical Debate on Urban Planning (Part III)**

- **The Ecology Argument (*cont'd*)**
  —Yet, for the critics, the ecology argument is flawed, not because it is not correct in pointing out the serious ecological problem challenging "our planetary survival system," but because the nature of sustainability is complex and thus uncertain in outcome.
  —To put it differently, "sustainability can only be determined in hindsight," because "we operate within the paradox of interdependence, requiring a predictable environment which, to retain its adaptive ability, must continually change." (M. Dempster 1998; R. Costanza 1996)
  —So, the ecology argument is naïve, since "[w]e can never know if current activities are sustainable, but can only infer their potential....These are complicated by recognition of the complex, unpredictable systems characteristics" in nature. (M. Dempster 1998)

- **The Fairness Argument**
  —Thirdly, a different (though related) theoretical approach can be called, in the absence of better terms, the *fairness* argument, in that any urban planning should consider the issue of social justice, especially for those who are disadvantaged, be they the poor, ethnic/racial minorities, social deviants, and others, who often have the misfortune to live in bad urban neighborhoods, while the rich enjoy nicer ones.
  —For instance, Jason Corburn (2008) argued for "an environmental justice framework that connects ecological, economic and social justice issues, especially in urban settings. Corburn asks about the distribution of environmental goods..., who participates in rule-making and enforcement, and how environmental justice evolves institutionally, and is enforced."
  —Phillip Thompson (2008), for another, spoke "for many lower-income city dwellers" when he asked, "Who, for example, will pay for outfitting buildings in poorer neighborhoods...and who will train and educate the workforce required for this transformation?"

*(continued on next page)*

**Table 1.2. The Theoretical Debate on Urban Planning (Part IV)**

- **The Fairness Argument (*cont'd*)**
  —Chris Zegras (2008) likewsie championed the need for the alleviation of the "fundamental issues of accessibility for the poor: their lack of affordable transportation and proximity to schools and jobs. Zegras recommends addressing the worldwide crisis in transportation, in part through such innovations as bike and car sharing."
  —And Adil Najam (2008) shared the same argument by observing that, if the earth is one single country, then it "is poor, extremely divided, degraded, poorly governed and unsafe—a Third-world country."
  —Yet, for the critics, the intention of all these scholars is noble indeed, fighting for social justice on behalf of the poor, the minorities, and other unfortunate ones.
  —But, the deeper question for them here concerns the age-old question which has divided philosophers through the ages, namely, What is justice? Who is going to say what is just or unjust, fair or unfair, so that every group on earth can agree?
  —In *The Future of Human Civilization* (*FHC*), I already argued why this search for social justice on earth is essentially conflicting, to the extent that there is no morality without immorality, no justice without injustice, or no equality without inequality.

- **The Complexity Argument**
  —Fourthly, an alternative theoretical approach can be referred to as the *complexity* argument, in that it takes seriously the "complexity" of the closed linkages between the natural environment and the human one in urban planning, in a way which tries to resolve the problem of uncertainty in outcome—unlike the ecological argument, which is rejected (by those in the complexity argument) as a bit too naïve, albeit noble in intention.

*(continued on next page)*

**Table 1.2. The Theoretical Debate on Urban Planning (Part V)**

- **The Complexity Argument (*cont'd*)**
  —The understanding of "complexity" is not an easy task. For instance, some scholars like Seth Miller go so far as to propose "at least 45 different definitions of 'complexity.'" (J. Rosser 2003) Others, however, like David Green (2001) and David Newth, identify six main features of complexity, namely, (i) "self-organization," (ii) "connectivity," (iii) "criticality," (iv) "novelty," (v) "diversity," and (vi) "emergence"—as this is something that I already went to great lengths to analyze in *The Future of Complexity* (*FC*).
  —With this in mind—it is not surprising that, while "it is critical to recognize fundamental differences between the 'engineering' paradigm and newer approaches based on complex systems concepts," the "[a]pplication of 'new' systems thinking (e.g. self-organization, complexity) to planning seems to be quite limited as yet in the formal literature." (M. Dempster 1998)
  —But the critics are quick to point out two essential limits in the complexity argument for resolving the issue of sustainability in urban planning.
  —Firstly, the first problem concerns "our current abilities and approaches. Can we cope with the complexities and uncertainties inherent in these systems?" (M. Dempster 1998) M. Beth Dempster (1998) thus confessed: "My general answer to these questions is—not adequately."
  —And secondly, the second problem is a follow-up to the first, in that if we now understand the uncertain nature of many complex systems, the challenge then is to come up with a way to address the issue of sustainability within a complex system more realistically.
  —But, for the critics, the challenge here remains: How exactly should urban planning be done in such a way as to deal with "the paradox of interdependence, requiring a predictable environment which, to retain its adaptive ability, must continually change," together with its uncertain characteristics in outcome which are typical in complex systems?

*(continued on next page)*

**Table 1.2. The Theoretical Debate on Urban Planning (Part VI)**

---

• **The Contingency Argument**

—And fifthly, unlike the previous three theoretical approaches—an alternative one can be called, in the absence of better words, the *contingency* argument, which learns from the previous four theoretical approaches without, however, favoring one argument over the other and, in the process, transcends them all.

—The contingency argument, in the end, has the virtue to show that the concern with sustainability has been much exaggerated and distorted, to the point that it is fast becoming a new intellectual fad, so that its dark sides have been unwarrantedly ignored or downgraded.

—Thus, the contingency argument is my original theoretical contribution to the debate on the issue of sustainability in the context of urban planning, with the analysis of "density" and "void" as a case study.

—My *contingency* argument can be more precisely called *the contingency theory of urban planning*, since it proposes a new theory of urban planning for the study of sustainability. This contingency theory of urban planning transcends the previous four arguments by adding something new to the literature, yet without indulging in the euphoric discourse on the possibility and desirability of sustainability.

—In other words, the contingency theory of urban planning has seven major theses, namely, (a) the first thesis: the denseness-emptiness principle, (b) the second thesis: the absoluteness-relativeness principle, (c) the third thesis: the simpleness-complicatedness principle, (d) the fourth thesis: the evolution-transformation principle, (e) the fifth thesis: the symmetry-asymmetry principle, (f) the sixth thesis: the regression-progression principle, and (g) the seventh thesis: the post-human planning—to be elaborated in the rest of the book, with a summary in the concluding chapter.

—If true, this seminal theory will fundamentally change the way that we think about the issue of sustainability by transcending it, with enormous implications not only for understanding the future of urban planning, in a small sense—but also for predicting the relevance of sustainability in relation to the entire domain of human knowledge for the human future and what I originally called its "post-human" fate, in a broad sense.

---

*Sources*: A summary of *Sec.1.4* in *FPHUP*

**Table 1.3. The Theoretical Debate on Space-Time (Part I)**

- **Isaac Newton's Absolutist (Substantivist) Theory of Space-Time**
  —space and time are independent from each other. The structure of space-time is $E3 \times E1$ (with the structure of space, *P*, as the set of spatial locations in a three-dimensional Euclidean space, *E3*, and the structure of time as the set of temporal moments, *T*, in the one-dimensional real time, *E1*).
  —space and time are also independent from the effects of matter and events. The existence of space and time is possible even in a world absent of matter (and, for that matter, even in a world absent of events), as if they were material objects but with their total unchangingness thorough time.

- **Albert Einstein's Relativist Theory of Space-Time**
  —space and time are interchangeable (not absolute), just as matter and energy are equivalent (not independent) with the famous equation, $E = mc^2$ (as in the special theory of relativity in 1905).
  —space-time and matter-energy are also relative in a grand union (as in the general theory of relativity in 1915). Thus, each pair affects the other pair, as "matter 'tells' spacetime how to curve [in a non-Euclidean geometry] and curved spacetime 'tells' matter how to behave….Space contracts near mass and dilates away from it. Time dilates near mass and contracts away from it. Clocks positioned farther away from the mass of the earth run faster than clocks closer to the earth." (L. Shlain 1991: 328-330)

*(continued on next page)*

**Table 1.3. The Theoretical Debate on Space-Time (Part II)**

- **Peter Baofu's Perspectival Theory of Space-Time**
  —space and time can be understood from multiple perspectives, be they in relation to culture, society, nature, and the mind, with each perspective revealing something about the nature of space-time and simultaneously delimiting its view. This is subject to "the regression-progression principle" in existential dialectics.
  —each perspective of space and time exists in society and culture with good reasons, with some being more successful and hegemonic (dominant) than others. This is subject to "the symmetry-asymmetry principle" in existential dialectics.
  —space and time will not last, to be eventually superseded (altered) by post-humans in different forms (e.g.,stretching/shrinking space-time, engineering more dimensions of space-time, and manipulating multiverses), be they here in this universe or in multiverses. Thus, even the physical existence of space-time cannot last forever, with ever more transformations in the process. This is subject to "the change-constancy principle" in existential dialectics.
  —the conventional wisdom (especially by physicists) of treating the physical perspective of space and time as the foundation of all other perspectives (of space and time) and of regarding them as much less important is a form of reductionism,committing what I call *the foundation fallacy*, in misleadingly dismissing the multiple perspectives of space and time in relation to culture, society, nature, and the mind.

*Notes*: The examples in each category are solely illustrative (not exhaustive), and the comparison is relative (not absolute), nor are they necessarily mutually exclusive. Some can be easily re-classified elsewhere. As generalities, they allow exceptions.
*Source*: A summary of *Sec.1.1*, *Sec.1.2*, and *Sec.1.3* of *FPHST*—and, for that matter, the rest of *FPHST*

**Table 1.4. Main Reasons for Altering Space-Time**

- **The Need to Make New Energy-Matter**
  —Ex: manipulating molecular bonds for new materials
  —Ex: creating nanotechnologies on the atomic scale
  —Ex: engineering the atomic nucleus
  —Ex: restructuring most elementary particles
  —Ex: inventing new forms of matter and energy

- **The Need to Create New Space-Time**
  —Ex: creating "warp drive" (as in science fiction) for space travel
  —Ex: creating "pocket universes"

- **The Need to Conquer the Cosmos unto Multiverses**
  —Ex: spreading floating consciousness and hyper-spatial consciousness, besides other forms that humans have never known, in the cosmos and beyond unto multiverses for ultimate conquest

*Notes*: The examples in each category are solely illustrative (not exhaustive) nor necessarily mutually exclusive, and the comparison is relative (not absolute). As generalities, they allow exceptions. Also, it should be stressed that the three reasons are all related, in that they all contribute to the evolution of intelligent life in the cosmos unto multiverses in the most distant future beyond our current knowledge.
*Sources*: A summary of *Sec.6.2* of *FPHST*. See also *FHC*, *FCD*, *FPHC*, *FC*, *FPHU*, and *FPHG*, for example.

**Table 1.5. The Technological Frontiers of the Micro-World**

- **Type I-Minus**
  —Ex: building structures and mining

- **Type II-Minus**
  —Ex: playing with the genetic makeups of living things

- **Type III-Minus**
  —Ex: manipulating molecular bonds for new materials

- **Type IV-Minus**
  —Ex: creating nanotechnologies on the atomic scale

- **Type V-Minus**
  —Ex: engineering the atomic nucleus

- **Type VI-Minus**
  —Ex: restructuring most elementary particles

- **Type Ω-Minus**
  —Ex: altering the structure of space-time

*Notes*: As already indicated in *Sec.4.4.2.2* of *FPHC*, the problem with this micro-classification (from Barrow's work) is that the civilization types (with the exception of Type Ω-Minus, for example) are not quite distinct, since many of them can be achieved more or less in a civilization, to the extent that Type II-minus and Type III-minus, just to cite two plausible types, can be historically contemporaneous, relatively speaking, unlike the vast historical distance between, say, Type O and Type I (or Type I and Type II) civilizations. In other words, the micro-classification here is not very useful to understand civilization types but is revealing to see the technological frontiers of the micro-world.
*Sources*: A reconstruction from J.Barrow (1998:133), as originally shown in *Table 4.7* of *FPHC*. See *FPHC* for more info.

**Table 1.6. Finity, Transfinity, and Infinity (Part I)**

• **Infinity**

—The term "infinity," which is "symbolically represented with ∞," derives from the Latin *infinitas* (or "unboundedness"), to refer to "several distinct concepts—usually linked to the idea of 'without end'—which arise in philosophy, mathematics, and theology." (WK 2008r)

—In the context of mathematics, "infinity" is often "treated as if it were a number (i.e., it counts or measures things: 'an infinite number of terms') but it is a different type of 'number' from the real numbers. Infinity is related to limits,…large cardinals,…projective geometry, extended real numbers and the absolute Infinite," for instance. (WK 2008r)

• **Transfinity**

—To avoid confusion, my usage of the word "transfinite" (as used here in the ontological principle) is not the same as the one which was originally coined by Georg Cantor to refer to "cardinal numbers or ordinal numbers that are larger than all finite numbers, yet not necessarily absolutely infinite" (in the mind of God), as opposed to "relative" infinity (in the mind of Man). (WK 2008s) In other words, for Cantor, there are two kinds of infinity, "absolute" and "relative"—with "relative" infinity or "transfinity" (in the mind of Man) being dependent on "absolute" infinity (in the mind of God) for its existence.

—By contrast, my usage of the word "transfinite" here differs radically from the one used by Cantor (and other mathematicians) and is more limited, in light of the problems confronting any attempt to understand the idea of infinity, be it by intuition, imagination, and conception (as already shown in the prevous three sub-sections).

—With this clarification in mind, I allow numbers which can be many times larger—or smaller, for that matter—than the finite things that we encounter in daily life, but they do not have to be related to the idea of infinity at all (which may not exist).

*(continued on next page)*

**Table 1.6. Finity, Transfinity, and Infinity (Part II)**

- **Transfinity** ***(cont'd)***
  —Of course, there may be some *borderline* cases, in which it is not clear whether the number in question is transfinite (in my usage) or simply a mathematical convenience. A good example of a borderline case is the Planck unit of length for "the smallest space possibly measured in nature," which is "less than billionths of trillionths of trillionths of an inch" (or something like $1.6 \times 10^{-35}$ meters). (P. Baofu 2006a; N. McAleer 1987: 219; WKV 2008; D. Corbett 2008)
  —The ideas concerning symmetry and dimensions in the examples above were addressed in Chapter Three and Chapter Four of *FPHG*. For now, it suffices to further add that it is mind-boggling to even conceive or imagine extreme numbers like the Planck unit of length (or others). Does it really exist at all? Or is it merely a mathematical convenience to describe a physical situation that the human mind does not understand properly?
  —In any event, "transfinity" (in my usage) can be used as a further extension (of number) to the finite numbers in daily life—albeit with some borderline cases in mind.

- **Finity**
  —It refers to numbers which are bounded (that is, with an end), especially (though not exclusively) in relation to things in everyday life.

*Source*: From *Sec. 1.4* and *Sec. 2.2.3* of *FPHG*

**Table 1.7. Theoretical Speculations of Multiverses**

- **"Baby Universes" (Ex: Andre Linde and others)**
  —Ex: In a flat universe theory, "even if our part of it eventually collapses,…some spots in the cosmos would suddenly start inflating on their own, creating brand-new 'baby universes.'" (P. Baofu 2000: 623)

- **"Parallel Universes" (Ex: Stephen Hawking and others)**
  —Ex: In quantum cosmology, there allows the existence of infinite numbers of parallel universes, with tunneling among them. (M. Kaku 1994: 256) Hawking later revised his views on this.

- **"Pocket Universes" (Ex: Alan Guth)**
  —Ex: "As the pocket universes live out their lives and recollapse or dwindle away, new universes are generated to take their place....While life in our pocket universe will presumably die out, life in the universe as a whole will thrive for eternity." (A. Guth 1997: 248; P. Baofu 2002: 482)

- **"Brane Worlds" (Ex: Warren Siegel, Lisa Randall, and others)**
  —Ex: Our universe is stuck on a membrane of space-time embedded in a larger cosmos, with different brane worlds connecting and/or colliding with each other.

*Notes*: These examples are solely illustrative (not exhaustive), and some of the items can be reclassified somewhere else. Nor are they always mutually exclusive. Since they are generalities, exceptions are expected.
*Source*: From *Table 4.8* of *FPHC*

**Table 1.8. The Confusion between "Many Worlds" and "Multiverse" (Part I)**

- **First Problem**
  —The first problem in this second confusion concerns a debate between the "many-worlds" interpretation of quantum mechanics by Hugh Everett (1957) and others, and the Copenhagen interpretation by Niels Bohr and Werner Heisenberg on the effects of quantum states after interacting with an external environment (e.g., by a measuring observer).
  —On the one hand, the Copenhagen interpretation of quantum mechanics is more indeterministic in its concern with the probabilities of the outcomes of a wave function after interacting with an external environment (e.g., by a measuring observer using light to detect, for example, "an electron in a particular region around the nucleus at a particular time." (WK 2008t) Each wave function for the state of a particle refers to "a mathematical representation used to calculate the probability for it to be found in a location, or state of motion." (WK 2008u) And the "wavefunction collapse" means that "the act of measurement causes the calculated set of probabilities to 'collapse' to the value defined by the measurement." (WK 2008u)
  —On the other hand, the Everett's interrpetation is more deterministic, in rejecting "the objective reality of wavefunction collapse" and "instead explaining the subjective appearance of wavefunction collapse with the mechanism of quantum decoherence." (WK 2008v) The words "quantum decoherence" here refers to "the mechanism by which quantum systems interact with their environments to exhibit probabilistically additive behavior," in such a way that "the quantum nature of the system is simply 'leaked' into the environment." (WK 2008v) For example, "[a]s a result of an interaction, the wave functions of the system and the measuring device become entangled with each other. Decoherence happens when different portions of the system's wavefunction become entangled *in different ways* [italic added] with the measuring device...." (WK 2008v) Thus, for Everett, the appearance of "wavefunction collapse" can be explained by "quantum decoherence" for quantum states to branch out in different independent worlds after interacting with an external environment.

*(continued on next page)*

**Table 1.8. The Confusion between "Many Worlds" and "Multiverse" (Part II)**

- **First Problem *(cont'd)***
  —Yet, a major criticism against Everett's interpretation is that these different independent worlds in Everett's sense "will never be accessible to us," because of the interference by the measuring observer in an external environment. (WK 2008v)
  —For illustration, "once a measurement is done, the measured system becomes entangled with both the physicist who measured it and a huge number of other particles, some of which are photons flying away towards the other end of the universe; in order to prove that the wave function did not collapse one would have to bring all these particles back and measure them again, together with the system that was measured originally. This is completely impractical, but even if one can theoretically do this, it would destroy any evidence that the original measurement took place (including the physicist's memory)." (WK 2008v)
  —But then, the critics ask, If these independent worlds in Everett's sense are inaccessible to us, how do we know much of anything really about them?

- **Second Problem**
  —What is relevant at the quantum mechanical level may not necessarily apply to the world at the physicsl level.
  —In fact, the persistent inability among contemporoary physicists to unifiy the theory of relativity for large objects at the physical level and quantum mechanics for tiny objects at the quantum mechanical level should warn us about the danger of reductionism, by reducing the understanding of different universes ("parallel universes") at the physical level from a generalization of "many worlds" at the quantum mechanical level.

- **Third Problem**
  —There is no empirical evidence to suggest that there are infinite (or extremely many) possible worlds in the cosmos as the many-worlds model by Everett would have us to believe.

*Source*: From *Sec.4.4.1* of *FPHG*

**Table 1.9. Hyperspace and Its Challenge (Part I)**

- **First Problem**
  —Hyperspace for intra- and inter-universal travel is assumed to be possible, on the basis of four main assumptions; although there may be others, these are the often cited ones.
  —The first assumption concerns the existence of "entry" and "exit" points in hyperspace, which is easier said than done to enter, go through, and then exit hyperspace.
  —For instance, using hyperspace "requires complex calculations…of single or multiple hyperspace jumps and the control of the jump….Peturbations such as those experienced by ship in space from the gravitional field around an object such as a planet or even a star are exacerbated in hyperspatial travel, since mass in real space distorts hyperspace in an equal measure. 'Jumping' near to a gravitational mass is likely to make resulting exit from hyperspace to be highly uncertain, with the level of improbability *i* increasing with the square of the distance to the nearest gravitional 'well.'" (WK 2008w)
  —Besides, "[o]nly spaceships equipped with a special force field can enter hyperspace, because exposure to hyperspace even for short period of time is hazardous to unprotected humans." (WK 2008w)

- **Second Problem**
  —The second assumption concerns the faster-than-light speed. It is well known that Einstein argued for the speed of light as the ultimate limit, but nowadays phyiscists do allow the violation of this limit under certain conditions (like the hypothetical existence of tachyons, as will be explained further in a later sub-section on time travel).
  —Even then, there is a challenge to explain in science fiction "why ships can travel faster than light in hyperspace," and two good arguments are that "hyperspace may be smaller than real space and therefore a star ship's propulsion seems to be greatly multiplied, or else the speed of light in hyperspace is not a barrier as it is in real space. Whatever the reasoning, the general effect is that ships traveling in hyperspace seem to have broken the speed of light, appearing at their destinations much more quickly and without the shift in time that the Theory of Relativity would suggest." (WK 2008w)

*(continued on next page)*

**Table 1.9. Hyperspace and Its Challenge (Part II)**

- **Third Problem**
  —The third assumption concerns the speed of time (to be measured, say, by a clock), but this is elaborated in a different section on the related issue of "time travel" (as shown in a different table on time travel).

- **Fourth Problem**
  —The fourth assumption concerns spatial shortcut, as it is possible, because of the curvature of space, in that "the manifold of ordinary three-dimensional space is curved in four or more 'higher' spacial dimensions (a 'hyperspace' in the geometric sense…). This curvature causes certain widely separated points in three-dimensional space to nonetheless be 'adjacent' to each other four-dimensionally. Creating an aperture in 4D space (a wormhole) between these locations can allow instantaneous transit between the two locations; a common comparison is that of a folded piece of paper, where a hole punched through two folded sections is more direct than a line drawn between them on the sheet. This idea probably arose out of certain popular descriptions of General Relativity and/or Riemannian manifolds, and may be the original form from which later concepts of hyperspace arose." (WK 2008w)
  —But this does not mean that hyperpsace is always safe. In fact, "in some science fiction, the danger of hyperspace travel is due to the chance that the route through hyperspace may take a ship too close to a celestial body with a large gravitational field, such as a star. In such scenarios, if a starship passes too close to a large gravitational field while in hyperspace, the ship is forcibly pulled out of hyperspace and reverts to normal space. Therefore, certain hyperspace 'routes' may be mapped out that are safe, not passing too close to stars or other dangers." (WJ 2008w)

*Source*: From *Sec.4.4.1* of *FPHG*

**Table 1.10. The Problems of Time Travel into the Future (Part I)**

- **First Problem**

—The first problem concerns "time dilation under the Theory of Special Relativity" (which separates space-time from mass-energy as special cases). Suppose you are on a moving train and an observer outside the train is watching you along the tracks. As your train moves along at the speed of light (or something close to it), "time, as measured by your watch, ticked along at a slower pace than time measured by the observer. Not only that, distance changed, too. For the observer, a one-foot ruler whizzing by on the train would have measured less than a foot." (NO 2000)

—In fact, you may not even notice that this slower clicking of the clock; in fact, everything looks normal to you inside the train as it was before: "The weird thing is that, for you on the train, time wouldn't seem to be moving slower and your ruler wouldn't be shorter—all would appear normal." (NO 2008) By contrast, you instead think that "time on the rest of the Earth would appear to be ticking along slower and its rulers would be shorter." (NO 2000) So, both of you insist that the other clock (not his) is clicking more slowly—assuming, of course, an inertial frame of reference.

—Now, change the story a bit, in that, this time, instead of an "inertial reference frame," acceleration is allowed, and you are on your way to board a spaceship. As an illustration, suppose "[y]ou board a spaceship and take off for deep space. The ship approaches the speed of light. Time for you seems to pass as it always has. It takes you about five seconds to tie your shoe. But to an observer on Earth (assuming he or she could watch you), you are moving at a snail's pace. It takes hours to tie your shoe." (NO 2000) In any event, "you continue on your journey. You slow down, stop, and accelerate back to Earth. You arrive home. You have aged two years during your flight. Two hundred years have passed on Earth. You have successfully travelled forward through time." (NO 2000) In this case, both the observer (if he is still alive) and you agree that you have aged less.

*(continued on next page)*

**Table 1.10. The Problems of Time Travel into the Future (Part II)**

• **First Problem** ***(cont'd)***

—In other words, when you are back to Earth, a lot of things have changed since, and you may find rather strange to be in a new environment which you hardly recognize, when compared with what you used to think as your city, or the Earth (two years ago, well, in accordance to your time of measurement). This means that you are in the future (two hundred years later), even though you are still relatively young (only two years older). But, can you go back? Well, no. The reason is that "[a]ccording to relativity, you can only move through time in one direction." (NO 2000)

• **Second Problem**

—The second problem concerns "time dilation under the Theory of General Relativity" (which unifies space-time with mass-energy in a general theory), the same result of time dilation occurs.

—For instance, when "one clock is deeper in a gravity well than the other," a result is that "the clock deeper in the well" will "tick…more slowly; this effect must be taken into account when calibrating the clocks on the satellites of the Global Positioning System, and it could lead to significant differences in rates of aging for observers at different distances from a black hole." (WK 2008x)

—More precisely, by way of another example, "it has been calculated that, under general relativity, a person could travel forward in time at a rate four times that of distant observers by residing inside a spherical shell with a diameter of 5 meters and the mass of Jupiter. For such a person, every one second of their 'personal' time would correspond to four seconds for distant observers. Of course, squeezing the mass of a large planet into such a structure is not expected to be within our technological capabilities in the near future." (WK 2008x; R. Gott 2002)

—The point here, however, is that time travel into the future, in this sense, is allowed, theoretically speaking, but in a very limited way, from the theory of relativity. But one can also ask, What if the theory of relativity were wrong? Well, in that case, there would be a need to find a better theory to explain time travel into the future.

*Source*: From *Sec.4.4.2* of *FPHG*

**Table 1.11. The Problems of Time Travel into the Past (Part I)**

- **First Problem**
  —The first problem concerns the faster-than-light-speed, since, in accordance to the theory of special relativity, when an object goes faster than the speed of light, something weird can happen, in that, instead of having the clock clicking more slowly, it is moving backward, in relation to some inertial frame of reference—which then raisies the issue of whether time travel into the past is possible at all.
  —But the problem here is about the violation of the law of casuality. In everyday language, an excellent illustration of causality violation is that, suppose you travel back into the past and then kill your parents, this then means that you could not have been born. But this is not true, since you are alive.
  —Yet, "in the case of a hypothetical signal moving faster than light, there would always be some frames in which the signal was received before it was sent, so that the signal could be said to have moved backwards in time." The violation of casuality exists in this case.
  —The first part of the problem is that, in accordance to the theory of special relativity, "it would take an infinite amount of energy to accelerate a slower-than-light object to the speed of light"—let alone the energy needed to propel an object to go faster than the speed of light. (WK 2008x) Although some suggest the possibility of "negative energy," it remains to be seen to what extent this is true.
  —And the second part of the problem is that, "although relativity does not forbid the theoretical possibility of tachyons which move faster than light at all times, when analyzed using quantum field theory it seems that it would not actually be possible to use them to transmit information faster than light and there is no evidence for their existence." (WK 2008x; S. Chase 1993)

*(continued on next page)*

**Table 1.11. The Problems of Time Travel into the Past (Part II)**

• **Second Problem**

—The second problem concerns cosmic string and black holes, since it is the theory of general relativity which "extends the special theory to cover gravity, illustrating it in terms of curvature in spacetime caused by mass-energy and the flow of momentum." (WK 2008x)

—On the other hand, "[g]eneral relativity describes the universe under a system of field equations…and there exist solutions to these equations that permit what are called 'closed time-like curves,' and hence time travel into the past….The first of these was proposed by Kurt Gödel, a solution known as the Gödel metric, but his (and many others') example requires the universe to have physical characteristics that it does not appear to have." (WK 2008x; K. Thorne 1994)

—But then, the problem now becomes: "Whether general relativity forbids closed time-like curves for all realistic conditions is unknown." (WK 2008x)

• **Third Problem**

—The third problem concerns wormholes and related ones, and "[w]ormholes are a hypothetical warped spacetime which are also permitted by the Einstein field equations of general relativity, although it would be impossible to travel through a wormhole unless it was what is known as a traversable wormhole." (WK 2008x; M. Visser 1995) In other words, "a wormhole is a hypothetical topological feature of spacetime that is fundamentally a 'shortcut' through space and time. Spacetime can be viewed as a 2D surface, and when 'folded' over, a wormhole bridge can be formed. A wormhole has at least two mouths which are connected to a single throat or tube. If the wormhole is traversable, matter can 'travel' from one mouth to the other by passing through the throat." (WK 2008y)

—But the problem here is three-fold, as shown in what follows.

*(continued on next page)*

**Table 1.11. The Problems of Time Travel into the Past (Part III)**

---

- **Third Problem *(cont'd)***
  —The first part of the problem is that "there is no observational evidence for wormholes." (WK 2008y)
  —The second part of the problem is that it is not clear how exactly wormholes (and black holes, for that matter) could allow time travel into the past at all.
  —And the third part of the problem is that if using wormholes (and black holes, for that matter) to travel into the past requires something to go faster than light too, then this raises another issue concerning the two problems as previously indicated in the sub-section on the faster-than-light-speed.

- **Fourth Problem**
  —The fourth problem is that some physicists have argued that, even should time travel into the past be possible, by way of these three methods, the person still would not return to the same past history that he exactly experienced before, but in a different one. (WK 2008x)
  —In other words, "strictly speaking," time travel into the past is not "really" possible, in light of these criticisms, so there is no need to worry that one could travel into the past and kill one's parent, as a way to violate the law of casuality.

---

*Source*: From *Sec.4.4.2* of *FPHG*

**Table 1.12. The Conceptual Dimensions of Consciousness (and Other Mental States)**

- **On Heredity and Time**
  —Heredity and the Environment
  —The Past and the Present

- **On Layers of Mental States and Abnormality**
  —Consciousness, Unconsciousness, and Preconsciousness
  —Normality and Paranormality

- **On Organicity and Motivation**
  —Mechanicity and Organicity
  —Primary Motivations and Multiple Motivations

- **On Other (Mostly Epistemic) Considerations**
  —Synthesis and Analysis
  —Situation and the Subject
  —Process and Outcome
  —Reasoning and Other Modes of Thinking
  —Meta-Conceptual Nominalism and Realism

*Notes*:: These examples are solely illustrative (not exhaustive), and some of the items can be reclassified somewhere else. Nor are they always mutually exclusive. Since they are generalities, exceptions are expected.
*Sources*: From *FPHC*. A re-construction, but with my own contribution, originally from G.Lindzey & C.Hall, *Introduction to Theories of Personality* (NY: John Wiley & Sons, 1985).

**Table 1.13. The Theoretical Levels of Consciousness (and Other Mental States) (Part I)**

- **At the Micro-Physical Theoretical Level**
  - —Quantum-Mechanics
  - —Electromagnetism

- **At the Chemical Theoretical Level**
  - —Biochemistry

- **At the Biological Theoretical Level**
  - —Evolutionary Biology
  - —Neuroscience
  - —Artificial Intelligence

- **At the Psychological Theoretical Level**
  - —Psychodynamic Psychology
    - Psychoanalytical Psychology
    - Analytical Psychology
    - Socially Oriented Psychology
  - —Experimental (Behavioral) Psychology
    - Operant Reinforcement Theory
    - Stimulus-Response Theory
    - Social Learning Theory
  - —Cognitive (Gestalt) Psychology
    - Humanist Psychology
    - Existential Psychology
    - Field Theory
  - —Psychometric Psychology
    - Idiographics
    - Constitutional Psychology
  - —Social Psychology
    - Symbolic Interactive Theory
    - Social Exchange Theory

*(continued on next page)*

**Table 1.13. The Theoretical Levels of Consciousness (and Other Mental States) (Part II)**

- **At the Organizational Theoretical Level**
  —Managerial-Bureaucratic Theory
  —Oligarchic Theory
  —Network Theory

- **At the Institutional Theoretical Level**
  —Functionalist Theory
  —Anomic Theory

- **At the Structural Theoretical Level**
  —Conflict Theory
    - Marxian Theory
    - Critical Theory
    - Weberian Theory
  —Games Theory (in Formal Theory)
  —Feminist Theory
    - Feminist Concerns
    - Feminist Hopes

- **At the Systemic Theoretical Level**
  —Equilibrium Theory
  —System Theory
  —Chaos Theory

- **At the Cosmological Theoretical Level**
  —Superluminal Model
  —The Theory of Floating Consciousness

*(continued on next page)*

**Table 1.13. The Theoretical Levels of Consciousness (and Other Mental States) (Part III)**

- **At the Cultural Theoretical Level**
  - —Substantive Theories
    - Structuralist Theory
    - Post-Structuralist Theory (in Postmodernism)
  - —Meta-Theories
    - Epistemic Objectivism vs. Epistemic Historicism
    - Epistemic Subjectivism vs. Epistemic Non-Subjectivism
      - –Phenomenology
      - –Ethnomethodology
      - –Hermeneutics
    - Epistemic Relativism vs. Epistemic Absolutism
    - Epistemic Reductionism vs. Epistemic Emergencism

- **At Other Levels**
  - —Structuration Theory
  - —Reflexive Socioanalysis

*Notes*: These examples are solely illustrative (not exhaustive), and some of the items can be reclassified somewhere else. Nor are they always mutually exclusive. Since they are generalities, exceptions are expected.
*Source*: From many different sources as indicated in *FPHC*

**Table 1.14. The Thematic Issues of Consciousness (and Other Mental States)**

---

- **The Factor of History**
  —Pre-Modernity
  —Modernity
  —Postmodernity
  —After-Postmodernity

- **The Importance of Needs**
  —Having (e.g., the everyday, the technological)
  —Belonging (e.g., the just and the good)
  —Being (e.g., the true, the holy, the sublime/beautiful, and the good)

---

*Notes*: These examples are solely illustrative (not exhaustive). Nor are they always mutually exclusive. Since they are generalities, exceptions are expected. Also, some categories overlap with others; for instance, the good is also related with being-ness, just as the everyday can be connected with belong-ness.

*Sources*: From *FPHC*, based on *FHC* and *FCD*

**Table 1.15. Having, Belonging, and Being in Consciousness**

- **Having (e.g., About the Everyday and Technological)**
  —Ex: Physiological needs (e.g., thirst, hunger, sex)
  —Ex: Safety (e.g., stability, freedom from fear and anxiety)

- **Belonging (e.g., About the Good and Just)**
  —Ex: Social interaction
  —Ex: Friendship, acquaintance
  —Ex: Love, family
  —Ex: Self-respect, respect from others

- **Being (e.g., About the True, Holy, Beautiful/Sublime, and Good)**
  —Ex: Understanding of reality about self and world
  —Ex: Spiritual quest for holiness, beauty, sublimity, and goodness

*Notes*: These examples are solely illustrative (not exhaustive), and some of the items can be reclassified somewhere else. Nor are they always mutually exclusive. Since they are generalities, exceptions are expected.
*Sources*: From *FPHC*, based on *FHC* and *FCD*

**Table 1.16. The Having-Ness of Consciousness (and Other Mental States)**

- **At the Micro-Physical Level**
  —Ex: The finer physics of nutrition

- **At the Chemical Level**
  —Ex: Food guide pyramid

- **At the Bio-Psychological Level**
  —Ex: The hierarchy of needs

- **At the Institutional Level**
  —Ex: The production of commodity fetishism

- **At the Organizational Level**
  —Ex: Mass standardization (and customization); bureaucratic rationality

- **At the Structural Level**
  —Ex: The divide between the haves and the have-nots

- **At the Cultural Level**
  —Ex: The bondage of greater expectations

- **At the Systemic Level**
  —Ex: The technological transformation of poverty with transvaluations

- **At the Cosmological Level**
  —Ex: Floating consciousness in deep space without the human physiological needs

*Notes*: These examples are solely illustrative (not exhaustive), and some of the items can be reclassified somewhere else. Nor are they always mutually exclusive. Since they are generalities, exceptions are expected.
*Sources*: From *FPHC*. See also *FCD* and *FHC*.

**Table 1.17. The Belonging-Ness of Consciousness (and Other Mental States)**

- **At the Micro-Physical Level**
  —Ex: Statistical physics and social networks

- **At the Chemical Level**
  —Ex: Olfaction and social relations

- **At the Bio-Psychological Level**
  —Ex: Dual human nature on relationships

- **At the Institutional Level**
  —Ex: Private property and social cleavage

- **At the Organizational Level**
  —Ex: Legal formalism and iron cage

- **At the Structural Level**
  —Ex: Class conflicts and social alienation

- **At the Cultural Level**
  —Ex: Ruthless society and the culture of law

- **At the Systemic Level**
  —Ex: Human extinction and post-human relations

- **At the Cosmological Level**
  —Ex: The transcendence of equality in the cosmos and beyond

*Notes*: These examples are solely illustrative (not exhaustive), and some of the items can be reclassified somewhere else. Nor are they always mutually exclusive. Since they are generalities, exceptions are expected.
*Sources*: From *FPHC*. See also *FCD* and *FHC*.

**Table 1.18. The Being-Ness of Consciousness (and Other Mental States) (Part I)**

- **At the Micro-Physical Level**
  —Ex: Space-time and matter-energy in classical mechanics, quantum mechanics, and the theory of relativity

- **At the Chemical Level**
  —Ex: Drugs and well being

- **At the Biological Level**
  —Ex: Neuroscience and mystical experiences

- **At the Psychological Level**
  —Ex: Reductionism and emergencism

- **At the Institutional Level**
  —Ex: The proliferation of cults and sects, and the corporate-art industry

- **At the Organizational Level**
  —Ex: The organization of ideas in the scheme of things

- **At the Structural Level**
  —Ex: Disciplinary power and control in the order of things

- **At the Systemic Level**
  —Ex: Mutualism and ecological psychology

- **At the Cultural Level**
  —Ex: The recurrence of competing spirits, and floating consciousness as a climax of evolution

*(continued on next page)*

**Table 1.18. The Being-Ness of Consciousness (and Other Mental States) (Part II)**

- **At the Cosmological Level**
  —Ex: The search for elsewherewhen, and the evolution of other consciousnesses (e.g., the hyper-spatial consciousness)

*Notes*: These examples are solely illustrative (not exhaustive), and some of the items can be reclassified somewhere else. Nor are they always mutually exclusive. Since they are generalities, exceptions are expected.
*Sources*: From *FPHC*. See also *FCD* and *FHC*.

**Table 1.19. Cognitive Partiality in Different Mental States**

- **The Conscious**
  - —*Biased*
    - Ex: Self-overratingness
    - Ex: Self-modesty
    - Ex: Self-fulfilling prophesy
  - —*Shallow*
    - Ex: Attribution error
    - Ex: Heuristics
    - Ex: Illusory thinking
  - —*Short-term*
    - Ex: Deindividuation
  - —*Materialistic*
    - Ex: Hierarchy of needs

- **The Subconscious**
    - Ex: mental reconstruction
    - Ex: stereotyping

- **The Unconscious**
    - Ex: automatic processing
    - Ex: cognitive intuition

*Notes*: These examples are solely illustrative (not exhaustive), and some of the items can be reclassified somewhere else. Nor are they always mutually exclusive. Since they are generalities, exceptions are expected. And the comparison is relative, not absolute.
*Source*: From *BCPC*

**Table 1.20. Emotional Non-Neutrality and Behavioral Alteration in Different Mental States**

- **The Emotional**
  - —*Conscious*
    - Ex: Feelings
  - —*Subconscious*
    - Ex: Moods
  - —*Unconscious*
    - Ex: Empathy
    - Ex: Instincts

- **The Behavioral**
  - —*Conscious*
    - Ex: Foot-in-the-door
    - Ex: Role playing
    - Ex: Saying-is-believing
  - —*Subconscious*
    - Ex: Discrimination
  - —*Unconscious*
    - Ex: Non-Verbal Communication

*Notes*: These examples are solely illustrative (not exhaustive), and some of the items can be reclassified somewhere else. Nor are they always mutually exclusive. Since they are generalities, exceptions are expected. And the comparison is relative, not absolute.
*Source*: From *BCPC*

**Table 1.21. The Limits of Intuition in Unconsciousness**

• **Cognitive Intuition (or Intuitive Thinking)**
—*Features*
Ex: mental schemas
Ex: expertise
Ex: blindsight
—*Problems*
Ex: inaccurate
Ex: non-explanatory

• **Emotional Intuition**
—*Features*
Ex: empathy
Ex: instincts
—*Problems*
Ex: over-generalized
Ex: non-explanatory

• **Behavioral Intuition**
—*Features*
Ex: prosemics
Ex: kinesics
—*Problems*
Ex: unreliable
Ex: non-explanatory

*Notes*: These examples are solely illustrative (not exhaustive), and some of the items can be reclassified somewhere else. Nor are they always mutually exclusive. Since they are generalities, exceptions are expected. And the comparison is relative, not absolute.
*Source*: From *BCPC*

**Table 1.22. The Wealth/Poverty Dialectics in Different Mental States: The Case of Cognition**

- **The Conscious**
  - —*Biased*
    - Ex: If richer in self-serving bias, then poorer in understanding others
  - —*Shallow*
    - Ex: If richer in heuristic thinking, then poorer in comprehending reality
  - —*Short-term*
    - Ex: If richer in short-term tendency, then poorer in long-term planning
  - —*Materialistic*
    - Ex: If richer in physiological obsession, then poorer in spiritual enlightenment

- **The Subconscious**
  - Ex: If richer in mental reconstruction, then poorer in accuracy of recall
  - Ex: If richer in stereotyping, then poorer in understanding groups

- **The Unconscious**
  - Ex: If richer in automatic processing, then poorer in depth of analysis
  - Ex: If richer in cognitive intuition, then poorer in reliability of judgments

*Notes*: These examples are solely illustrative (not exhaustive), and some of the items can be reclassified somewhere else. Nor are they always mutually exclusive. Since they are generalities, exceptions are expected. And the comparison is relative, not absolute.
*Source*: From *BCPC*

**Table 1.23. The Wealth/Poverty Dialectics in Different Mental States: The Case of Emotion and Behavior**

- **Emotion**
  - —*Conscious*
    - Ex: If richer in relying on familiarity and similarity for social attraction, then poorer in having diverse social relations
  - —*Subconscious*
    - Ex: If richer in moodiness, then poorer in sober thinking
  - —*Unconscious*
    - Ex: If richer in empathy, then poorer in task-oriented efficiency
    - Ex: If richer in instincts, then poorer in critical thinking

- **Behavior**
  - —*Conscious*
    - Ex: If richer in foot-in-the-door technique, then poorer in instant Gratification
    - Ex: If richer in role playing, then poorer in identity stability
  - —*Subconscious*
    - Ex: If richer in discriminating, then poorer in social harmony
  - —*Unconscious*
    - Ex: If richer in kinesics and proxemics, then poorer in verbal skills, relatively speaking.

*Notes*: These examples are solely illustrative (not exhaustive), and some of the items can be reclassified somewhere else. Nor are they always mutually exclusive. Since they are generalities, exceptions are expected. And the comparison is relative, not absolute.
*Source*: From *BCPC*

**Table 1.24. The Theoretical Debate on Nature vs. Nurture (Part I)**

---

- **The Environmental Approach**
  —*Thesis*: It focuses, relatively speaking, more on the environment (culture and society) in explaining the achievement gap among individuals and for that matter, countries or regions, when contrasted with the natural factors.
  —*Discourse*: Examples include Jose Ortega y Gasset ("Man has no nature; what he has is history"), Ashley Montagu ("Man is man because he has no instincts, because everything he is and has become he has learned from his culture, from the man-made part of the environment, from other human beings"), Stephen Jay Gould ("[The] brain [is] capable of a full range of behaviors and predisposed to none"), and Jesse Jackson (who blames white racism for the failure of blacks to close the achievement gap between whites and blacks in America). The works on dependency theory in international political economy (with a Marxian influence) and on the Protestant work ethic (by Max Weber) also point to this environmental direction.

- **The Genetic Approach**
  —*Thesis*: It focuses instead, relatively speaking again, on hereditory factors (e.g., genes and evolution) in explaining the achievement gap among individuals and for that matter, countries or regions, when contrasted with the envrionmental factors.
  —*Discourse*: Examples are Hans Eysenck and William Sheldon (in constitutional psychology), Konrad Lorenz (in his work on innate aggressive human nature), Gary Marcus (on the complexities of human thought by a tiny number of genes) and Robert Plomin, together with Michael Owen and Peter McGuffin (on the genetic basis of complex human behaviors).

---

*(continued on next page)*

**Table 1.24. The Theoretical Debate on Nature vs. Nurture (Part II)**

- **The Compromise Approach**
  —*Thesis*: It seeks the middle-of-the-road argument in regard to nature and nurture and regards all differences among individuals and groups as the result of the mixture of both nature and nurture, more or less equally.
  —*Discourse*: Examples include C. Murray and R. Herrnstein ("It seems highly likely to us that both genes and environment have something to do with this issue") and Dan Dennett ("Surely 'everyone knows' that the nature-nurture debate was resolved long ago, and neither side wins since everything – is – a – mixture – of – both – and - it's – all – very -complicated…."

- **The Transcendent Approach**
  —*Thesis*: It goes beyond both nature and nurture (without, however, committing the compromise fallacy) in showing their closely intertwined interactions in producing the behavioral differences as often seen in individual human endeavors on the micro scale, and for that matter, in country (or regional) endeavors on the macro one—in the context of my five theses, namely, (a) the compromise fallacy, (b) no oppression without self-oppression, (c) no success without failure, (d) the factor of randomness, and (e) the post-human vision, to be elaborated in Chapter Six.
  —*Discourse*: Peter Baofu proposed this approach on the basis of his "theory of contrastive advantages" (as an original theoretical contribution to the debate, which was first proposed and analyzed in *The Future of Capitalism and Democracy*). In the end, the human genes will not last, to be eventually superseded by post-human life forms, so the debate between genes and memes has obscured something profoundly important about the future that the world has never known. And the debate is also misleading and faulty in its dichotomy.

*Source*: From *BNN*. See the book for citations and details.

**Table 1.25. Physical Challenges to Hyper-Spatial Consciousness**

- **The Understanding of a Higher-Dimensional World of Space-Time**
  —Ex: 4 for traditional aspects of space-time (e.g., length, width, breadth and time) plus 6 more new dimensions in theory of hyper-space, with profound implications for practical applications to new forms of consciousness.

- **The Mastering of Dark Matter and Dark Energy**
  —Ex: "ordinary matter" (e.g., atoms, molecules) as a mere 4.4% of the universe, with 23% made of "cold dark matter" and the rest (about 73%) of mysterious "dark energy," with fundamental significance to questions about the limit of the speed of energy (or info), the availability of energy for use, and the nature of space-time, just to cite some examples.

- **The Exploration of Multiverses**
  —Ex: theoretical speculation of other universes (e.g., "baby universes," "gateways" in black holes, "wave function of the universe," "many worlds," "brane worlds"), with potentially seminary discoveries of different physical laws in relation to matter-energy and space-time, and vital differences to the future of post-human conquest of other universes (for the emergence of new forms of consciousness).

*Notes*: These examples are solely illustrative (not exhaustive), and some of the items can be reclassified somewhere else. Nor are they always mutually exclusive. Since they are generalities, exceptions are expected. The point here is to give a rough picture of the evolution of consciousness to the hyper-spatial consciousness and others totally unknown to current earthlings. As a note of clarification, it makes no difference to my argument as to whether or not the hyper-spatial consciousness may emerge before, during, and after floating consciousness.
*Source*: From *Table 4.5* of *FPHC*

**Table 1.26. The Theory of Floating Consciousness (Part I)**

---

- **At the Micro-Physical Level**
  —Ex: intelligent life without the human physical-chemical system

- **At the Chemical Level**
  —Ex: space radiation and toxins

- **At the Bio-Psychological Level**
  —Ex: exo-biological evolution in deep space
  —Ex: genetic engineering of new beings

- **At the Institutional Level**
  —Ex: post-capitalism
  —Ex: post-democracy

- **At the Organizational Level**
  —Ex: less legal-formalistic routines

- **At the Structural Level**
  —Ex: alien forms of violence

- **At the Cultural Level**
  —Ex: transcending freedom
  —Ex: transcending equality

---

*(continued on next page)*

**Table 1.26. The Theory of Floating Consciousness (Part II)**

---

• **At the Cosmological Level**
—Ex: parallel universes
—Ex: pocket universes

• **At the Systemic Level**
—Ex: space habitats (in zero-gravity)

---

*Notes*: Each example draws from the works of different scholars in the field. For instance, at the cosmological level, the idea of parallel universes is from the theoretical speculation in quantum cosmology by Stephen Hawking and others, while the one of pocket universes comes from the theoretical work of Allan Guth at MIT. And at the institutional level, I proposed post-capitalism and post-democracy in *FCD* (and later, from *BDPD* and *BCPC*). In addition, the examples are solely illustrative (not exhaustive), and some of the items can be reclassified somewhere else. Nor are they always mutually exclusive. Since they are generalities, exceptions are expected.
*Sources*: From *FPHC*—and, originally, from *FCD*

**Table 1.27. The Potential of Unfolding Unconsciousness**

- **Superior Senses**
  - —*Through space*
    - Ex: clair-sentience (in feeling)
    - Ex: clair-voyance (in seeing)
    - Ex: clair-audience (in hearing)
    - Ex: clair-austance (in tasting)
    - Ex: clair-alience (in smelling)
    - Ex: clair-cognizance (in knowing)
  - —*Through time*
    - Ex: pre-cognition (in knowing), pre-sentiment (in feeling)
    - Ex: retro-cognition (in knowing), retro-sentiment (in feeling)

- **Superior Contacts**
  - —*With living minds*
    - Ex: telepathy
  - —*With the dead*
    - Ex: mediumship
  - —*With other-living entities*
    - Ex: telekinesis

*Notes*: The categories and examples are solely illustrative (not exhaustive), and the comparison is relative (not absolute), nor are they necessarily mutually exclusive. And some can be easily re-classified elsewhere. As generalities, they allow exceptions.
*Source*: A summary of Ch.5 of *FPHU*

**Table 1.28. The Future Exploration of Unfolding Unconsciousness**

- **In the Context of Nature**
  - —Ex: the higher-dimensional space-time in hyperspace theory
  - —Ex: the particle-wave interactions in quantum mechanics
  - —Ex: the study of brain waves in electromagnetic theory

- **In the Context of the Mind**
  - —Ex: the unconscious fantasies in psychoanalysis
  - —Ex: the mind-altering drugs in biochemistry
  - —Ex: the manipulation of neural activity in neurobiology

- **In the Context of Culture**
  - —Ex: the correlation between anomalous experience and moral/religious/aesthetic interests in culture studies
  - —Ex: the role of myths and language in epistemology

- **In the Context of Society**
  - —Ex: different organizational agendas in anomalous research
  - —Ex: the scholarly divide on anomalous study
  - —Ex: interests and propaganda in institutioanl research
  - —Ex: the transformative power of technology in research on social systems
  - —Ex: the natural and non-natural selection of other worlds in exo-biology
  - —Ex: the role of the post-human conscious mind in qualitative demography

*Notes*: The examples in each category are solely illustrative (not exhaustive), and the comparison is relative (not absolute), nor are they necessarily mutually exclusive. And some can be easily re-classified elsewhere. As generalities, they allow exceptions.
*Source*: A summary of Chs.2-5 of *FPHU*, in conjunction with the rest of the book

**Table 1.29. Creative Techniques and Traits**

- **Creative Techniques**
  - —*Comprehensiveness*
    - Ex: Broadness of Scope
    - Ex: In-Depth Analysis
    - Ex: Multifaceted Taxonomy
    - Ex: Numerous Theories
  - —*Arrangement*
    - Ex: Separation (or Subtraction)
    - Ex: Combination (or Addition)
    - Ex: Replacement
    - Ex: Reversal
    - Ex: Accommodation
    - Ex: Stretching
  - —*Serendipity*
    - Ex: Play
    - Ex: Chance

- **Creative Traits**
  - —*Openness / Risk*
    - Ex: Ability to welcome new ideas or to do new things, with risk
  - —*Discipline*
    - Ex: Ability to sacrifice whatever it takes to succeed
  - —*Resilience / Confidence*
    - Ex: Ability to take failures to mind, not to heart
  - —*Trust*
    - Ex: Ability to cooperate with others for team work, if necessary

*Notes*: The examples in the categories are solely illustrative (not exhaustive), and the comparison is relative (not absolute), nor are they necessarily mutually exclusive. And some can be easily re-classified elsewhere. As generalities, they allow exceptions.

*Source*: From *Sec.4.2* of *FPHCT*. For more details, also read the rest of the book.

**Table 1.30. The Desirability of Creativity, and Its Dark Sides (Part I)**

- **In Relation to Instrumental Rationality**
  - —*Desirability*
    - Ex: For the sake of efficiency and effectiveness in this capitalist era of our time.
  - —*Undesirability*
    - Ex: What Max Weber (1930) called the "iron cage" of capitalism, where live the "sensualists without heart, specialists without spirit"—or something which the Frankfurt School has forcefully asked us to be freed from.

- **In Relation to Substantive Rationality**
  - —*Desirability*
    - Ex: For the sake of God, the King, Motherland, or other comparable ideals.
  - —*Undesirability*
    - Ex: Domination and oppression of various forms—something which the moderns had struggled to free themselves from in the first place, since the dawn of modernity.

- **In Relation to Autonomous Rationality**
  - —*Desirability*
    - Ex: For the sake of itself, the autonomy of creative endeavor.
  - —*Undesirability*
    - Ex: The decadence of its own degeneration (e.g., the donwsides of postmodernism in our postmodern times, or what Nietzsche once called the unbearable "unreality" and "falseness" of the autonomous artist's "innermost existence"—and, in other cases, the painful suffering from different mental illnesses for those highly creative individuals).

*(continued on next page)*

**Table 1.30. The Desirability of Creativity, and Its Dark Sides (Part II)**

*Notes*: The examples in the categories are solely illustrative (not exhaustive), and the comparison is relative (not absolute), nor are they necessarily mutually exclusive. And some can be easily re-classified elsewhere. As generalities, they allow exceptions.
Also, I already went to great lenghts to explain the desirability and its dark sides of all these three rationalities (although there may be others too) in the two volumes of *FHC*—and, for that matter, *FCD*.
*Source*: From *Sec.4.9* of *FPHCT*. For more details, also read the rest of the book.

**Table 1.31. Posthuman-Ism, Post-Humanism, and Trans-Humanism**

- **Post-Humanism**
  - —The neologism "post-human" used in my books should *not* be confused with another term which looks similar but has a totally different meaning in the literature of postmodernism, namely, "post-humanism"—which constitutes a critique of "humanism" as tradionally understood (especially, though not exclusively, in relation to the idea of progress in science and reason in the Enlightenment project). (WK 2008)
  - —My works reject the project of "postmodernism" and propose the future world of what I orignally called "after-postmodernity" in *FHC* and *FCD*, for instance.

- **Trans-Humanism**
  - —Also, the neologism "post-human" used in my books should *not* be confused with a similar term which is used to champion the ideology of technology for the fturue co-existence between humans and cyborgs in "trans-humanism." (WK 2008a)
  - —Instead, my term "post-human" in relation to "posthuman-ism" also rejects "trans-humanism" (especially, though not exclusively, in relation to the promises of technology) and refers to something else altogether, that is, the future extinction of humans and its post-human successors in deep space and beyond unto multiverses.

- **Posthuman-Ism**
  - —With these two clarifications in mind, the "post-humans" (as envisioned in my books) can take different forms, and I proposed, in my different books, some of them, such as "unfolding unconsciousness," "floating consciousness," "hyper-spatial consciousness," "thinking machines," "thinking robots," "genetically altered superior beings," "cyborgs," and others (as already analyzed in my previous works).
  - —They will evolve in conjunction with other visions of mine in relation to nature, society, and culture.

*Source*: From *Sec.1.6* of *FPHE*. For more details, also read the rest of the book (and other books of mine).

**Table 1.32. The Trinity of Pre-Modernity**

- **Pre-Free-Spirited Pre-Modernity (Pre-Modernism) and Its Internal Split**
  —Competing worldviews and values both within and between linear centric (e.g., Islamic, Christian, Judaic, Imperial Roman) and cyclical-centric (e.g., Confucian, Taoist, Hindu, and Buddhist) orientations
  —Compare modernism with pre-modernism here in relation to the seven dimensions of human existence like the true and the holy (e.g., different versions of epistemic dogmas and religious superstitions), the everyday and the technological (e.g.,different versions of non-technophilism and non-consumerism), the beautiful/sublime (e.g., different versions of aesthetic non-autonomy), and the good and the just (e.g., different versions of moral particularism).

- **Pre-Capitalist Pre-Modernity (Pre-Modernization) and Its Own Discontents**
  —Competing versions of societal arrangements (e.g., feudalism, monarchism, and the holy order)

- **Hegemonic Pre-Modernity and Its Countervailing Forces**
  —Different power centers and their enemies (e.g., the Roman Empire and the "barbarian hordes," the "Holy Crusades" and the Muslims, the Middle Kingdom and the invading tribes, different social castes in India, and warring Greek city-states)

*Notes*: The examples in each category are solely illustrative (not exhaustive) nor necessarily mutually exclusive, and the comparison is relative (not absolute). As generalities, they allow exceptions. Also, it does not matter what the "base" era is in the analysis of any trinity. And in the present context, the "base" era is modernity (for instance, with its "free-spirited modernity" and the other two parts). So, for pre-modernity, the trinity takes the form of, say, "pre-free-spirited pre-modernity," together with the other two parts.
*Sources*: From Ch.2 of *BCIV* and also the 2 volumes of *FHC*

**Table 1.33. The Trinity of Modernity (Part I)**

- **Free-Spirited Modernity (Modernism) and Its Internal Split**
  - —*On the True and the Holy*
    - The freedom from the dogmas of the past to the better understanding of, and union with, the world and self (Ch.3 of *FHC*)
    - Alternative discourses: about the true (e.g., anti-science discourses) and the holy (non-mainstream theologies) (Ch.3 of *FHC*)
  - —*On the Technological and the Everyday*
    - The freedom from life harshness to the higher technophilic, consumeristic lifeform (Ch.2 of *FHC*)
    - Alternative discourses: about the everyday (e.g., transcendental mindsets) and the technological (e.g., Arcadianism) (Ch.2 of *FHC*)
  - —*On the Good and the Just*
    - The freedom from the theo-aristocratic tyranny to the moral universality for a just society (Ch.5 of *FHC*)
    - Alternative discourses: about the just (e.g., Communism, Anarchism) and the good (e.g., Nazism/Fascism, and Zarathustrianism) (Chs.5-6 of *FHC*)
  - —*On the Beautiful and the Sublime*
    - The freedom from the external distortion of aesthetic pleasure to the boundless infinity of totality in artistic autonomy (Ch.4 of *FHC*)
    - Alternative discourses: about the beautiful/sublime (e.g., kitsch and historical avant-gardism) (Ch.4 of *FHC*)

- **Capitalist Modernity (Modernization) and Its Own Discontents**
  - —*During the Industrial Revolution*
    - Ex: Marx on the institution of inequality (Ch.1 of *FHC*)
  - —*During the Modern Rational-Instrumental Epoch*
    - Ex: Weber on the politics of soft liberal institutions (Ch.5 of *FHC*)
  - —*During the Great Depression*
    - Ex: Keynes on the myth of the free market (Chs.1-3 of *FHC*)
  - —*During the Cold War*
    - Ex: Lasch on the narcissistic culture industry (Chs.2-3 of *FHC*)

*(continued on next page)*

**Table 1.33. The Trinity of Modernity (Part II)**

- **Hegemonic Modernity and Its Countervailing Forces**

—*The Legacies of Colonialism and Imperialism*

- Ex: European colonization of most of the modern world (Ch.1 of *FHC*)

—*The Struggle for Decolonialization*

- Ex: The countervailing forces of resentment, rechantment, and regionalism (Chs.1 & 6 of *FHC*)

*Notes*: The examples in each category are solely illustrative (not exhaustive) nor necessarily mutually exclusive, and the comparison is relative (not absolute). As generalities, they allow exceptions.
*Sources*: From the 2 volumes of *FHC*—and also from *FCD*

**Table 1.34. The Trinity of Postmodernity**

- **Free-Spirited Postmodernity (Postmodernism) and Its Internal Split**
  —*On the True and the Holy*
    - Postmodern performative turn for knowing and its enemies (Ch.3 of *FHC*)
    - Postmodern comparative theology and its opponents (Ch.3 of *FHC*)
  —*On the Technological and the Everyday*
    - Postmodern corporate technological mindset and its adversaries (Ch.2 of *FHC*)
    - Postmodern postmaterialism and its critics (Ch.2 of *FHC*)
  —*On the Good and the Just*
    - Postmodern politics of difference and its foes (Ch.5 of *FHC*)
  —*On the Beautiful and the Sublime*
    - Postmodern deconstruction and its dissenters (Ch.4 of *FHC*)

- **Capitalist Postmodernity (Postmodernization) and Its Own Discontents**
  —*During the Post-Cold War and Beyond*
    - Ex: post-Fordism and its shortcomings (Ch.6 of *FHC*; Chs.6-7 of *FCD*)

- **Hegemonic Postmodernity and Its Countervailing Forces**
  —*The Debate on the Global Village*
    - Ex: uni-civilizationalism vs. multi-civilizationalism (Ch.6 of *FHC*)
  —*The Resistance Movement*
    - Ex: rechantment and the politics of civilizational claims (e.g., Islamic, Confucian and other ethos in relation to the Same) (Ch.6 of *FHC*; Ch.10 of *FCD*)
    - Ex: resentment and the politics of resurgence (e.g., the rising Chinese superpower, the growing EU, and other players in relation to the U.S. and her allies) (Ch.6 of *FHC*; Ch.8 of *FCD*)
    - Ex: regionalism and the politics of inequality (e.g., trans- or international blocs, the North-South divide, NGO's) (Ch.6 of *FHC*; Ch.5 of *FCD*)

*Notes*: The examples in each category are solely illustrative (not exhaustive) nor necessarily mutually exclusive, and the comparison is relative (not absolute). As generalities, they allow exceptions.
*Sources*: From *FCD* and the 2 volumes of *FHC*

**Table 1.35. The Trinity of After-Postmodernity**

- **Free-Spirited After-Postmodernity (After-Postmodernism) and Its Internal Split**
  —The discourse of naked contingency (Ch.10 of *FCD*; Ch.4 of *FPHC*)

- **Post-Capitalist After-Postmodernity (After-Postmodernization) and Its Own Discontents**
  —Different versions of post-capitalism and post-democracy, and their enemies (Ch.10 of *FCD*; Chs.3-4 of *FPHC*)

- **Hegemonic After-Postmodernity and Its Countervailing Forces**
  —The Cyclical Progression of Hegemony in Multiverses (Chs.9-10 of *FCD*; Ch.4 of *FPHC*)

*Notes*: The examples in each category are solely illustrative (not exhaustive) nor necessarily mutually exclusive, and the comparison is relative (not absolute). As generalities, they allow exceptions.
*Sources*: From *FCD* and also *FHC*

# • Part Two •

## *Density*

# CHAPTER 2
## *DENSITY AND ITS CROWDEDNESS*

[W]e must never forget that we are talking about our planetary survival system, not just some artifact that we, the human species, now suddenly consider important.
—Kevin McNamee (1996)

## The Function of Density

The issue of sustainability can be analyzed within the context of density in urban planning, especially when it is to address the questions of desirability and possibility of sustainability.

A good way to provide a comprehensive analysis of density in this sense is an examination from the four perspectives of the mind, nature, society, and culture.

In other words, this chapter is organized into four sections, in relation to (2.2) density and the mind, (2.3) density and nature, (2.4) density and society, and (2.5) density and culture—to be discussed respectively, in that order, in what follows (and summarized in *Table 2.1*).

## Density and the Mind

The nature of density in urban planning can be illuminated from the perspective of the mind, and two good case studies are in order here, namely, (2.2.1) traffic density, road rage, and human emotion; and (2.2.2) urban density and public health—to be discussed hereafter, respectively.

### Traffic Density, Road Rage, and Human Emotion

One form of traffic density is "traffic congestion," which refers to "a condition on any network as use increases and is characterized by slower speeds, longer trip times, and increased queueing. The most common example is the physical use of roads by vehicles." (WK 2008k)

In the case of the physical form of traffic congestion by vehicles, this can occur, because "[w]hen traffic demand is great enough that the interaction between vehicles slows the speed of the traffic stream, congestion is incurred. As demand approaches the capacity of a road (or of the intersections along the road), extreme traffic congestion sets in. When vehicles are fully stopped for periods of time, this is colloquially known as a traffic jam." (WK 2008k)

What then are the causes of traffic congestion? At first glance, some usual suspects can be easily identified, such as "sheer weight of traffic," "traffic incidents, road works and weather events." (WK 2008k)

Be so as it may, a more interesting question to ask concerns some possible solutions to traffic congestion, which can be divided into two parts for analysis, that is, both the demand side (i.e., "reducing traffic") and the supply side (i.e., "increasing road capacity"), in order to understand the nature of density, this time, in relation to traffic congestion.

Or, in the reverse direction, increasing traffic and decreasing road capacity can contribute to more traffic congestion over time.

*The Demand Side*

The reduction of traffic can be accomplished in different ways, and good examples include the following (which should be treated as being illustrative only, not exhaustive, and is summarized in *Table 2.2*): (WK 2008k)

- "Parking restrictions, making motor vehicle use less attractive by increasing the monetary and non-monetary costs of parking, introducing greater competition for limited city or road space. (D. Shoup 2005; H. Knoflacher 2006)
- "Park and ride facilities allowing parking at a distance and allowing continuation by public transport or ride sharing. Park-and-ride car parks are commonly found at metro stations, freeway entrances in suburban areas, and at the edge of smaller cities."
- "Reduction of road capacity to force traffic onto other travel modes. Methods include traffic calming and the shared space concept."
- "Road pricing, charging money for access onto a road/specific area at certain times, congestion levels or for certain road users"
- "Road space rationing, where regulatory restrictions prevent certain types of vehicles from driving under certain circumstances or in

certain areas," with the use of "number plate restrictions" and "permits." (LEDA 2008; J. Henley 2005)

Surely, the list can be much longer than this, but it has the virtue to show some of the factors which can affect the density of traffic.

*The Supply Side*

On the other side of the same coin, the increase (or decrease) of road capacity can affect the outcome of traffic density. Some good examples of road capacity increase include the following (and are summarized in *Table 2.2*): (WK 2008k)

- "Adding more capacity at bottlenecks (such as by adding more lanes at the expense of hard shoulders or safety zones, or by removing local obstacles like bridge supports and widening tunnels)"
- "Adding more capacity over the whole of a route (generally by adding more lanes)"
- "Creating new routes"

Again, these few examples are for illustration only, and of course there can be many other features to be cited.

But then, an interesting question to ask here then is, What are the consequences of traffic density?

*Consequences of Traffic Density*

To answer this question, it is important to remember that, while the above examples are sufficient to illustrate the increase or decrease of traffic congestion in relation to its demand and supply sides, traffic congestion can become a form of traffic density which has some serious effects.

Some good examples of these serious effects include the following (and are summarized in *Table 2.3*): (WK 2008k)

- "Wasting time of motorists and passengers ('opportunity cost'). As a non-productive activity for most people, congestion reduces regional economic health."
- "Delays, which may result in late arrival for employment, meetings, and education, resulting in lost business, disciplinary action or other personal losses."

- "Inability to forecast travel time accurately, leading to drivers allocating more time to travel 'just in case,' and less time on productive activities."
- "Wasted fuel increases air pollution and carbon dioxide emissions (which may contribute to global warming) owing to increased idling, acceleration and braking. Increased fuel use may also in theory cause a rise in fuel costs."
- "Wear and tear on vehicles as a result of idling in traffic and frequent acceleration and braking, leading to more frequent repairs and replacements."
- "Stressed and frustrated motorists, encouraging road rage and reduced health of motorists."

Even though these serious effects are also illustrative, they raise the important question concerning the possibility and desirability of sustainability with traffic density as a case study.

Since the effects can be many, one good way to manage the analysis is to provide a further elaboration of an item on the list for illustration.

*The Case of Road Rage*

Consider, say, the phenomenon of "road rage" in the last item on the list (in the previous paragraph) for further elaboration.

The term "road rage" refers to "the behavior by a driver of an automobile or other motor vehicle, which thus causes collisions or incidents on roadways." (WK 2008l)

Road rage can have a combination of the following characteristics: (WK 2008l)

- "Generally aggressive driving, including sudden acceleration, braking, and close tailgating."
- "Cutting others off in a lane, or deliberately preventing someone from merging."
- "Sounding the vehicle's horn or flashing lights excessively."
- "Rude gestures (such as 'the finger')."
- "Shouting verbal abuse, obscenities, or threats."
- "Intentionally causing a collision between vehicles."
- "Exiting the car to attempt to start a confrontation, including striking someone else's vehicle with an object."
- "Threatening to use or using a firearm or other deadly weapon."
- "Throwing projectiles from a moving vehicle with the intent of damaging other vehicles."

Although the characteristics are by no means exhaustive, they give some indication of the nature of road rage.

The important point to remember here is that road rage is a chronic problem which, in spite of numerous attempts to curb it, has not disappeared. But why?

For instance, in the U.S., already "14 U.S. states have passed laws against aggressive driving." (WK 2008l) Yet, the problem remains, in that, "[i]n the U.S., more than 300 cases of road rage annually have ended with serious injuries or even fatalities—1200 incidents per year, according to the AAA Foundation study, and rising yearly throughout the six years of the study that examined police records nationally." (WK 2008l)

*The Bio-Psychological Explanations*

A natural question to ask is therefore whether or not, or to what extent, road rage reflects certain aspects of human nature—be they about cognition, emotion, and behavior.

There is no lack of theories to explain aggressive behavior like road rage in traffic density. Consider, say, three illustrative psychological approaches below.

Firstly, a psychological approach is to regard it as a form of mental disorder. In fact, "[a]s early as 1997, therapists in the United States were working to certify road rage as a medical condition. It is already an official mental disorder in the Diagnostic and Statistical Manual of Mental Disorders. According to an article published by the Associated Press in June 2006, the behaviors typically associated with road rage are the result of intermittent explosive disorder. This conclusion was drawn from surveys of some 9,200 adults in the United States between 2001 and 2003 and was funded by the National Institute of Mental Health." (WK 2008l)

However, for the critics, this psychological view needs to be treated with caution. Just consider two problems here. (WK 2008l) The first problem is that it is not clear what the causes of road rage (as an "intermittent explosive disorder") really are. And the second problem is that it is also not clear if there is a general agreement on whether or not, or to what extent, "road rage" is really "a mental issue."

Secondly, another theoretical approach concerns Sigmund Freud's idea of the "death drive" (*Thanatos*), as part of human nature torn between "two conflicting central desires: the life drive (*libido*)" for "survival, propagation, hunger, thirst, and sex"—and "the death drive (*Thanatos*)" for the return to "an inorganic or dead state." (WK 2008m)

In the case of "thanatos," the energy "is known as anticathexis, represented an urge inherent in all living things to return to a state of calm:

in other words, an inorganic or dead state. Freud recognized Thanatos only in his later years and developed his theory on the death drive in *Beyond the Pleasure Principle*. Freud approached the paradox between the life drives and the death drives by defining pleasure and unpleasure." (WK 2008m)

In other words, "unpleasure refers to stimulus that the body receives....For example,...the bombardment of visual stimuli amidst rush hour traffic produces anxiety....Conversely, pleasure is a result of a decrease in stimuli (for example, a calm environment the body enters after having been subjected to a hectic environment). If pleasure increases as stimuli decreases, then the ultimate experience of pleasure for Freud would be zero stimulus, or death." (WK 2008m)

In this way, "the death drive functions simultaneously toward extreme pleasure, which leads to death. Freud addressed the conceptual dualities of pleasure and unpleasure, as well as sex/life and death, in his discussions on masochism and sadomasochism. The tension between Eros and Thanatos represented a revolution in his manner of thinking." (WK 2008m)

Thus, there can be some kind of "sadism" (or even, conversely, "masochism") in the occurrence of road rage, under certain psychoanalytic conditions. The individual who engages in road rage may find certain pleasure in inflicting pain on others (and, for that matter, on hurting himself), under certain conditions.

However, for the critics, the Freudian theory is problematic. For instance, H. J. Eysenck (1986) argued that "Freud...consistently mis-diagnosed his patients, fraudulently misrepresented case histories and that 'what is true in Freud is not new and what is new in Freud is not true.'" (WK 2008m)

Mikkel Borch-Jacobsen (2000) was equally critical, when he wrote that "Freud...did not hesitate to build grand theories on...non-existent foundations...[that] he disguised fragments of his self-analysis as 'objective' cases, that he concealed his sources, that he conveniently antedated some of his analyses,...that he inflated his therapeutic successes, that he slandered his opponents."

But this is not to say that Freud has no supporters. In fact, over the years, "the field of neuro-psychoanalysis has shown strong support for Freud's theories among neuroscientists and researchers such as Oliver Sacks [1984], Mark Solms [2002], Jaak Panksepp [1998], Douglas Watt, António Damásio [1994], Eric Kandel, Joseph E. LeDoux[1996]," and K. Kaplan-Solms (2000), "by pointing out brain structures relating to Freud's concepts such as libido, drives, the unconscious, and repression." (WK 2008m & 2008n)

And thirdly, another theoretical approach concerns the bias of human cognition and the distortion of human emotions, or what I originally called "the theory of cognitive partiality" (in *Sec.2.4* of *The Future of Capitalism and Democracy* or *FCD*, and *Sec.4.5.1.1* of *Beyond Capitalism to Post-Capitalism*, or *BCPC*) and "the theory of emotional non-neutrality" (in *Sec.2.5* of *FCD* and *Sec.4.5.2* of of *BCPC*), as summarized in *Table 1.19* and *Table 1.20*.

Without repeating what was already analyzed in these books, it suffices to merely note here that, more often than conventionally understood, individuals can be biased, shallow, and short-term in their everyday thinking process, just as they can be moved by the subjective feelings, needs, and instincts of their emotional state.

Other scholars like Cass Sunstein, Richard Thaler (2008), and Dan Ariely also accounted for the human "tendency to self-destruct" under certain conditions, by they due to "apathy, peer pressure, and the tendency to misperceive in predictable ways," for instance. (T. Dokoupil 2009)

And the well-known psychologist Phillip Zimbardo (2007) also showed in his work on situational psychology how and why even good individuals, under certain conditions of life stress, can end up behaving badly towards others. This is what he called "the Lucifer Effect," in his book with the same title. (P. Zimbardo 2007a)

In the current context of road rage, those who get involved in it can do so without completely understanding and accepting the subsequent and long-term consequences of their outburst. So, it is too rash to degrade them without taking into account of the bio-psychological nature of human cognition, emotion, and behavior.

Of course, the critics may say that human nature is not written in stone and is essentially socially and culturaly constructed, so to speak, and can be altered for different purposes. In fact, I already extensively anlaysed this theoretical debate on nature versus nurture in *Beyond Nature and Nurture* (*BNN*) and showed how this extreme view on social and cultural constructivism is highly problematic—which at times may even lead to some kinds of authoritarian utopianism.

In any event, the important point to remember here is that one needs to have a more realistic expectation in curbing road rage, as a case study of the possibility and desirability of sustainability in urban planning.

After all, human nature has both good and dark sides to be considered, and this is something that I also already went to great lengths to explain in *The Future of Post-Human Consciousness* (*FPHC*), as shown in *Table 2.4*.

*The Deeper Implication*

In the end, the debate here is by no means pointless, even though it is not completely settled. After all, any theoretical debate on human bio-psychological conditions which is completely settled is rather rare in history, if it ever exists. But this does not mean that one should therefore deny any talk of human nature, as Steven Pinter (2002a) warned us against this modern denial of human nature in his book titled *The Black Slate* (2002).

The deeper implication of the debate, however, is to raise an unsettling issue concerning the possiblity and desirability of sustainability, even if the analysis of road rage in the context of traffic density is used as a case study here.

In other words, does the very advocation for sustainability presuppose too much the good sides of human nature without seriously accounting for its dark sides too, such that sustainability is not as much possible and desirable as its proponents would like us to believe?

With the case study of road rage in the context of traffic density as an example, urban planners should think twice, in a broader context, about the unsettling state of affairs that urban planning may not be as much possible to be designed to be sustainable in the long run as they would like us to believe, because of some bio-psychological conditions in humans which predispose them, even if not absolutely, to behave in the way that they have—and, for that matter, may not be as much desirable as the urban planners would like, because of some forms of pleasure or, in general, characteristics inherent in the dark sides of human nature. Of course, there are other causes too, in relation to nature, society, and culture (as will be elaborated in the rest of the chapter).

In any event, this is not to suggest, needless to say, that road rage should be tolerated and even encouraged, such that drivers can do whatever they like, in the name of the dark sides of human nature. The point here is to lower the euphoric expectations often seen in the debate for sustainability, in that there exists the other dimension of human nature, which may render any sustainable urban planning problematic in a way that the proponents for sustainability may not want to admit. After all, why should we expect anything in life to last forever, as things often grow and later die, and can be made to renew, grow, and later die again, so continues the cycle? It is sobering to admit that any form of urban planning for sustainability is no exception either.

This debate can also teach us something important for the ontological principles in existential dialectics, and good examples include the absoluteness-relativeness principle, the explicability-inexplicability principle, the denseness-emptiness principle, the simpleness-complicatedness, the sloweness-quickness principle, the order-chaos principle, the evolution-transformation principle, the softness-hardness principle, and the regression-progression principle.

For instance, in relation to the absoluteness-relativeness principle, if there is absolute viewpoint (e.g., the focus on the good side of human nature for moral intergenerational equity, that is, to act now for the future persistent existence of posterity), there is likewise relative concern (e.g., the more balanced analysis of both good and dark sides of human nature, such as the issues concerning the short-term, biased, and shallow thinking process and the non-neutrality of human emotions, so that what is valid from one standpoint may not so from another).

In relation to the explicability-inexplicability principle, if there is explicability (e.g., a decrease of road capacity can lead to the increase of traffic density), there is inexplicability (e.g., the unsettled issue concerning the bio-psychological causes of road rage).

In relation to the denseness-emptiness principle, if there is denseness (e.g., the requirement of urban density for large-scale infrastructure to be built), there is emptiness (e.g., the occurrence of road rage as a personal attempt to fight traffic density for a less croweded traffic space).

In relation to the simpleness-complicatedness principle, if there is simple analysis (e.g., the simplistic analysis of the correlation between traffic density and road rage), there is also its complicated counterpart (e.g., the deeper issue concerning the bio-psychological causes of road rage).

In relation to the slowness-quickness principle, if there is quickness (e.g., faster traffic in roads with greater capacity), there is slowness too (e.g., the occurrence of slower traffic in cases of traffic congestion).

In relation to the order-chaos principle, if there is order (e.g., the construction of roads and the formulation of traffic regulations for orderly traffic), there is also chaos (e.g., the occurrence of traffic jam and accidents).

In relation to the evolution-transformation principle, if there is natural evolution (e.g., the physical ability of humans to walk or run), there is likewise the technological transformation (e.g., the technical invention of vehicles for humans to move faster).

In relation to the softness-hardness principle, if there is softness (e.g., the legal enforcement against road rage as a way to promote public safety and traffic civility), there is hardness too (e.g., the aggressive form of driving in road rage at the expense of other vehicles).

And in relation to the regression-progression principle, if there is progress (e.g., the construction of roads for faster movement from one place to another), there is its regression (e.g., the occurrence of traffic jam and road rage).

## Urban Density and Public Health

There is another way to look into the relationship between density and the mind—this time, in relation to urban density and public health, as an illustration.

*The Rise of Urban Population*

As a start, there is the definitional problem of how to correctly measure the term "urban." In the United States, for instance, "the Census Bureau considers any area with over 2,500 population to be urban," but the problem here is that "this is a 'population center' many people are more likely to call a village or hamlet than a city" instead. (K. Gebbie 2008)

A more common definition, therefore, is to define an urban area as one with "100,000 [population] or more, with a nucleus of at least 50,000 and surrounding communities that share a high degree of social and economic integration. By this definition, only the 276 major metropolitan areas of the United States are considered urban." (K. Gebbie 2008)

This definitional issue aside—our time stands out from other ones, in that there has been a steady rise of urban population over time. For instance, "the population of the world has become much more urban. In 1900, 39.6 percent of the United States population was defined as 'urban' by the Bureau of the Census; by 1990 the proportion considered urban was 75.2 percent." (K. Gebbie 2008)

But this "75.2 percent" is only an average, since there are myriad urban areas with varied demographic sizes. For instance, "in central New York City (Manhattan), there are 52,419 people per square mile; in Cook County, Illinois (Chicago), 5,398; in Los Angeles County, California, 2,183; and in Dade County, Florida (Miami), 996.1. In contrast, the overall United States population density is 70.3 people per square mile; and in 1790 the nation had a population density of 4.5 people per square mile." (K. Gebbie 2008)

*The Twins of Social Cohesion and Social Breakdown*

With the rise of urban population, urban space becomes denser over time, and in the process, there emerge two social problems on either side of the urban horizon, namely, "social cohesion" and "social breakdown."

Social cohesion occurs, in that, "[t]he greater the cohesion, the more likely the group is to work together, to share common values, and to find positive solutions to problems in ways that are inclusive of all members of the group." (K. Gebbie 2008)

On the other hand, social breakdown occurs, in that "individuals are left to struggle with the challenges of living alone, people turn on one another in ways that are damaging, and problems accumulate to a level incompatible with a healthy life." (K. Gebbie 2008)

In fact, even when individuals try to escape social breakdown by moving into less populated areas (or less dense urban space), this often creates new problems of a different kind, such as "the difficulties of organizing public transportation. Lack of mass transportation may mean increased pollution from individual use of internal combustion engines, and it may mean that individuals lacking a personal car may have difficulty reaching health services." (K. Gebbie 2008)

An interesting phenomenon to notice here is that both social cohesion and social breakdown often occurs together in major urban spaces, in that some "urban areas…have within them neighborhoods that could be described as fully cohesive, but far too many urban areas are at, or are closer to, the other end of the spectrum." (K. Gebbie 2008)

*The Dilemma of Public Health*

The twins of social cohesion and social breakdown have implications for the nature of public health, be it (a) in a dense urban space or (b) in a less dense one.

(a) On the one hand, in a dense urban space, with "a large number of people," there is "the risk of health and illness problems. The closer proximity and higher rate of face-to-face contact has a direct impact on the rates of transmissible diseases such as tuberculosis and other respiratory infections. It is no surprise that the resurgence of tuberculosis experienced in the United States in the late 1980s and early 1990s began in New York." (K. Gebbie 2008)

In fact, "[t]he high population density, and the use of large, poorly ventilated spaces as overnight sleeping accommodations for the homeless provided an ideal environment for the transmission of the bacillus. The fact that the public health resources were being strained by the arrival of another condition, HIV/AIDS (human immunodeficiency virus/acquired immunodeficiency syndrome), compounded the problem and meant that drug-resistant organisms were being shared. Health concerns as much as concerns for recreation space have been involved in the development of at least limited open spaces such as parks within concentrated urban areas." (K. Gebbie 2008)

In addition, "[t]he concentration of populations in urban areas also means an increased accumulation of waste products. Removal of human waste and garbage is a major commitment in any city, whether the mechanism chosen is completely public or funded by a mixture of public and private resources. The treatment of human waste is costly, and new requirements that protect both people and the environment from contaminants has meant a steady investment in upgrading treatment facilities and building new ones." (K. Gebbie 2008)

Thus, living in a dense urban space is more likely to experience "stress" of various kinds, which affects negatively the quality of life, especially when linked to the related problem of violent behavior.

In other words, "in addition to the difficulties related to the increased numbers of encounters, there is an increased level of stress, which is known to increase the risk of illness. The density of urban populations and the associated stresses have also been associated with increased rates of violence and, in the second half of the twentieth century, an increase in crime associated with an increase in the distribution, sale, and use of illegal drugs. Some of the crime directly involved the drug distribution networks, as they competed with one another for turf; other crimes were committed by those who became addicted as they attempted to find the resources to support their addictions. For example, one occupational risk that has been studied is the risk of violence to convenience store employees. Of 1,835 robberies of convenience stores in eastern metropolitan areas in 1992 and 1993, 63 percent involved the use of a firearm, and 12 percent were associated with an injury to at least one employee. All five reported fatalities were firearm-related." (K. Gebbie 2008)

(b) On the other hand, in a less dense urban space, with a smaller number of people, as in a suburban area, there are problems of a different kind, although the immediate benefits are easy to notice, as "there may be more open spaces for recreation or sport, and access to more remote areas is simpler." (K. Gebbie 2008)

The tradeoff here, however, is that "suburbs mean more widely dispersed individual homes, each needing access to utilities and transportation, and constructed in such a way that neighborhood cohesion may be difficult or impossible to develop." (K. Gebbie 2008)

The problem is compounded further with "the availability of individual automobile transport," which "in the United States has undoubtedly contributed to suburban sprawl, as have issues of social discrimination. These areas have also grown because of what has been labeled 'urban flight': the movement out of cities of the more affluent as new waves of immigrants, often from different ethnic or racial groups, moved in." (K. Gebbie 2008)

One often discussed consequence of "urban flight" is the cycle of urban decay, in that "[t]he apparent cost of maintaining or advancing a standard of living within the urban core was seen as too great. This flight, however, leaves older housing stock to be occupied by those of lower income levels, with less generation of taxes to support services, and the beginning of a downward spiral. When combined with the movement of industry because of restrictions on pollution, search for a cheaper labor pool, or simple displacement due to competition from elsewhere, the result can be a severe, area-wide depression. The cities of the so-called Rust Belt of the northeastern United States provide many vivid examples of this cycle." (K. Gebbie 2008)

*A More Balanced Conclusioin*

So, what conclusion is there to be drawn, if the twins of social cohesion and social breakdown occur, especially when contextualized within a dense urban space or, conversely, a less dense one?

Kristine Gebbie (2008) had this to conclude: "Urban health is a complex web of both threats to health and supports to health. It cannot be understood apart from an appreciation for the size and density of the populations involved, and it continues to evolve as economic shift, technological advances, and public expectations develop. But the likelihood that urban areas will continue to concentrate people, problems, and opportunities makes its health concerns unique and important."

So, this illustrative analysis of urban density and public health has the virtue to show the dilemma confronting urban planners, in that any urban planning, however well intended and designed, is likely to experience the double faces of social cohesion and social breakdown, and this makes little difference to whether or not an urban area is highly dense or less so, although the threats and supports to health do not occur in the same way in either case.

The nature of sustainability, in other words, has both costs and benefits, just as it is not as possible as conventionally understood, as all urban areas often go through the cycle of growth and decay. After all, nothing really lasts forever.

This analysis can also teach us something about the ontological principles in existential dialectics, and good examples include the absoluteness-relativeness principle, the denseness-emptiness principle, the change-constancy principle, the expansion-contraction principle, the evolution-transformation principle, the symmetry-asymmetry principle, the softness-hardness principle, and the regression-progression principle.

For instance, in relation to the absoluteness-relativeness principle, if there is absolute analysis (e.g., the sole concern with the positive aspects of urban density), there is also relative counterpart (e.g., the twin aspects of social cohesion and social breakdown in urban density, such that what is valid from one standpoint may not be so from another).

In relation to the denseness-emptiness principle, if there is density (e.g., the need of urban density for a large population), there is also emptiness (e.g., the need of less dense urban space for urban flight).

In relation the change-constancy principle, if there is change (e.g., the ever new evolution of "economic shift, technological advances, and public expectations" in relation to urban density), there is constancy (e.g., the persistent existence of social cohesion and social breakdown over time).

In relation to the expansion-contraction principle, if there is expansion (e.g., the expansion of an urban area into its suburban space), there is contraction (e.g., the occurrence of urban decay in the aftermath of urban flight).

In relation to the evolution-transformation principle, if there is evolution (e.g., the natural evolution of humans to move around), there is transformation (e.g., the technological construction of a metropolitan area for urban density).

In relation to the symmetry-asymmetry principle, if there is symmetry (e.g., the co-existence of different groups living together in an urban space), there is asymmetry (e.g., the possibility of urban flight for those with higher income levels, but the trap of urban slums for those with lower income levels).

In relation to the softness-hardness principle, if there is softness (e.g, the nice civility to live together in social cohesion in certain areas of urban density), there is also hardness (e.g, the occurrence of violent, criminal behaviors by some individuals against others in urban areas).

And in relation to the regression-progression principle, if there is progress (e.g., the development of metropolitan areas to accommodate the rise of urban population over time), there is regression (e.g., the dark side of urbanization, such as urban decay and social breakdown).

# Density and Nature

Another way to examine the issue of density in relation to sustainability in urban planning can be had from the perspective of nature.

A good case study is provided hereafter, namely, in the context of ecological complexity and the urban balance with nature.

## Ecological Complexity, and the Urban Balance with Nature

In urban planning, a serious issue concerns how to balance the need to use resources in nature and the impact of exploiting them.

So an important question to ask here is, What then is the tipping point, in order to sustain long-term urban planning?

*Finding the Tipping Point*

For instance, there is a school of thought in urban planning called "Intelligent Urbanism," which argues that "the balance with nature emphasizes the distinction between utilizing resources and exploiting them. It focuses on the thresholds beyond which deforestation, soil erosion, aquifer depletion, siltation and flooding reinforce one another in urban development, saving or destroying life support systems." (WK 2008o)

This concern with nature is to "promote…environmental assessments to identify fragile zones, threatened ecosystems and habitats that can be enhanced through conservation, density control, land use planning and open space design…. This…promotes life cycle building energy consumption and pollutant emission analysis." (WK 2008o; I. McHarg 1975)

The focus on life cycle is to understand the nature of "environmental equilibrium," such that "there is a level of human habitation intensity wherein the resources that are consumed will be replaced through the replenishing natural cycles of the seasons, creating environmental equilibrium. Embedded in the principle is contention that so long as nature can resurge each year; so long as the biomass can survive within its own eco-system; so long as the breeding grounds of fauna and avifauna are safe; so long as there is no erosion and the biomass is maintained, nature is only being utilized." (WK 2008o)

In other words, there is then a tipping point, or "a fragile line that is crossed when the fauna, which cross-fertilizes the flora, which sustains the soil, which supports the hillsides, is no longer there. Erosion, siltation of drainage networks and flooding result. After a point of no return, utilization of natural resources will outpace the natural ability of the eco-system to replenish itself. From there on degradation accelerates and amplifies. Deforestation, desertification, erosion, floods, fires and landslides all increase." (WK 2008o)

Some good examples of crossing the "fragile line" (or the "tipping point") of "ecological interdependence" include such "blatant 'acts against nature'" as "cutting of hillside trees, quarrying on slopes, dumping sewage and industrial waste into the natural drainage system, paving and plinthing excessively, and construction on steep slopes. This urban theory proposes that the urban ecological balance can be maintained when fragile areas are reserved, conservation of eco-systems is pursued, and low intensity habitation precincts are thoughtfully identified. Thus, the principles operate within the balance of nature, with a goal of protecting and conserving those elements of the ecology that nurture the environment." (WK 2008o)

*Key Challenges*

But it is easier said than done. After all, is the balance with nature really possible? In other words, what are the key challenges to be overcome to achieve the balance with nature? And is this balance even desirable, in the first place? In other words, are the costs too high to achieve the balance with nature?

An excellent illustration to analyze the key challenges to balance with nature in urban planning is to explore the role of "local ecological knowledge" (LEK) in the process of implementing sustainable urban planning.

Vesa Yli-Pelkonen and Johanna Kohl (2005), using the experiment with "local ecological knowledge" (LEK) in contemporary Finnish urban planning as a case study, identified four major key challenges to be overcome. (J. Paldanius 1997)

The four key challenges are, namely, (a) "who is able to participate," (b) "what kind of attitudes and resources (knowledge, skills, and time) for participation do stakeholders (residents, associations, and land use planners) have," (c) what "are the right participatory methods used in relation to plan situations and participation objectives," and (d) "how is local participation integrated into the land use planning and decisionmaking process." (J. Kohl 2005)

However, there are different (conflicting) models of how these key challenges are to be overcome.

*Conflicting Models*

For instance, M. Callon (1999) identified three models on sustainable urban planning in the context of gathering and using "local ecological knowledge" (LEK).

These three models are called, respectively, (a) "the public education model (M1)…based on the irreducible opposition between scientific and popular knowledge," (b) "the public debate model (M2), in which lay people have different levels of knowledge and can to some extent participate in information production," and (c) "the co-production of knowledge model (M3), in which 'lay people are actively involved in the creation of knowledge concerning them.'" (J. Kohl 2005)

One can instantly see that the three models reveal the lack of a general agreement on sustainable urban planning. A summary of them below suffices to illustrate this essential point.

(a) Firstly, in accordance to "the public education model (M1)"—lay people are more or less excluded, because of their perceived lack of deep understanding of the scientific basis of ecology and urban planning.

In other words, for those in favor of M1, "the subjective LEK of lay people has nothing to contribute to objective scientific ecological information on which planning and decisionmaking is almost solely based….Due to the prevailing mistrust between lay people and science, and in order to reduce this mistrust, and thus reduce critical opinions…and reach rational decisionmaking, scientific experts have to intensify educational and informative actions towards lay people." (J. Kohl 2005)

(b) Secondly, in accordance to "the public debate model (M2)"—lay people are allowed to participate in sustainable urban planning, but with limits.

In other wrods, for those in favor of M2, "local or lay knowledge, gathered by inquiries and public hearings from different actors or groups of actors, can enrich scientific expert information…,which forms a basis for planning and decisionmaking…, but even at its best is incomplete." (J. Kohl 2005)

In fact, this model can be further divided between M2a and M2b, to further delineate "the different forms of participation." (J. Kohl 2004 & 2005)

For example, in "M2a (called the public hearing model), planners hear local knowledge directly and genuinely from participants, and participation can influence the planning process. M2a illustrates the aim of the new participatory policy and, if this model is accurate, participation is working as the policymakers expected. However, even if participants are heard, there is a great risk that the participation still has a little influence on the final plan." (J. Kohl 2005)

And in "M2b (called the public representation model), local knowledge is heard, but indirectly through representatives of local participants. These can be, for instance, local resident or nature associations. Local knowledge has the potential to influence the planning process as well as in M2a, but the knowledge users have to remember that they are hearing the interpretation of the representatives. However, these representatives can potentially present the knowledge to users better than individual participants." (J. Kohl 2005)

And (c), thirdly, in accordance to "the co-production of knowledge model (M3)"—lay people can participate the most, in that "lay people have an essential role in the collective production of knowledge and scientific information." (J. Kohl 2005)

In other words, for those in favor of M3, "[w]ith regards to planning and decision making, LEK has become equally as strong as formal science…, or, in particular circumstances, has become more important….This could happen, for instance, when scientific ecological information is scarce and LEK is perceived as an essential part of combined information production." (J. Kohl 2005); J. Weber 2001)

*The Difficulties to Obtain Local Ecological Knowledge*

The conflicting three models reveal something controversial, in that people simply disagree on how (a) to obtain and (b) to use "local ecological knowledge" (LEK) for sustainable urban planning.

Let's consider first (a), that is, how to obtain "local ecological knowledge" (LEK), and then (b), that is, the issue about how to use LEK in the next sub-section.

For illustration, there are four difficulties here on (a), that is, how to collect LEK from lay people (as summarized in *Table 2.5*).

Firstly, a main difficulty here concerns "timing," in that a master plan is often already decided well before LEK has a chance to influence it. This then complicates the process of collecting LEK on time.

In other words, "residents and nature enthusiasts often think that influencing the plan with their opinions and knowledge is rather difficult. One challenge seems to be timing; when a sketch of a new plan is introduced to the public, a certain number of basic conditions have already been determined, such as the master plan, aim of surface area to be built, and architectural vision. The planners are judicially obliged to consider the participants' information, but if reservations have already been made regarding the construction, the new information cannot affect the plan proposal or the final implementation.... According to the legislation's spirit, participants should be viewed as partners in the planning process; however participants sometimes perceive themselves as competitors with planners. But as was mentioned, even if planners would like to incorporate LEK, their hands can be tied by earlier decisions." (J. Kohl 2005)

Secondly, another main difficulty concerns the scope of collecting LEK from all participants involved.

In other words, the difficulty here is that "[p]lanning officials inform the public concerning the availability of plan preparation material, but these information channels do not always reach all interested parties. Local residents, and other participants, may feel that the release of information is inadequate....Participants have to keep their eyes and ears open, or they may read in the newspaper that the plan has just been approved. Participation requires devotion, tenacity, and interest in urban development." (J. Kohl 2005)

Thirdly, a third difficulty concerns the use of public discussion in collecting LEK.

In other words, the difficulty with the use of public discussion is that "[p]ublic discussions are aimed at all interested parties, but it is rather uncommon that all participants are able to attend the forums at the same time....[S]ome participants perceive that other methods of influencing the plan are more useful than public discussions. The success of the plan workshops depends largely on the kind of individuals that represent the local people and on their motives and values....Furthermore, LEK can be obtained actively by means of questionnaires, which, as emphasized by the interviewed planners, need to be very clearly written. Moreover, it is not a given that the answers cover a wide enough array of residents to provide balanced information. The loudest opponents may well get their voices heard, while those who are satisfied stay silent. When considering all the obstacles, people may feel that influencing the process is very difficult, and if their opinions do not count much, they may stop participating." (J. Kohl 2005)

And fourthly, the even more serious difficulty concerns the enormous problem of trying to understand the "complexity" of nature, especially in relation to (i) "self-organization," (ii) "connectivity," (iii) "criticality," (iv) "novelty," (v) "diversity," and (vi) "emergence," to the effect that both order and chaos are part of the nature of things. This is something that I already went to great lengths to analyze in *The Future of Complexity* (*FC*).

In fact, the "complexity" argument in *Sec. 1.4* precisely points to this enormous challenge of understanding nature, given the "complexities" and "uncertainties" inherent in it. So, the challenge here is, How can individuals collect all the info as needed, if the ecosystem in question is more complex and uncertain than what the enthusiastic urban planners would like to admit?

But these four difficulties of how to collect LEK is only one side of the issue. Another side concerns the difficulties of how to use LEK.

*The Difficulties to Use Local Ecological Knowledge*

There are three main difficulties here about the use of local ecoological knowledge (LEK) for sustainable urban planning (as summarized in T*able 2.6*).

Firstly, a main difficulty here concerns the numerous (often conflicting) pieces of information on LEK to choose from. So, which ones are to be used and which ones are to be discarded? As Dennis Frenchman (2008) once observed, an recurrent challenging problem in designing and building cities is, Who are to decide what to include or exclude in the designing and building process?

In other words, "[p]lanners receive a variety of information and opinions through public hearings, workshops, written statements, and other methods of communication. However, it is challenging to analyze the information in order to use it in the plan. Received information regarding local nature may vary from opinions, feelings, and experiences to very knowledgeable species observations and well-founded perspectives on the development of urban green areas." (J. Kohl 2005)

Secondly, another main difficulty concerns the reliability of info on LEK. For instance, "[p]lanners prefer well-founded and reliable knowledge, but they must be able to determine how reliable the knowledge source is....The interviewed planners emphasize that local knowledge must be treated critically, because it may easily contain inaccuracies. In the case of species or biotope observations, planners principally trust LEK sources, and this information can be verified on location if necessary. The situation becomes more difficult if the observation is several years old, because circumstances may have changed. However, an old observation may indicate that the area is of potential importance for biodiversity. If the local green spaces are obliterated or altered, the species lose their potential habitat." (J. Kohl 2005)

And thirdly, a related difficulty concerns the subjective (or emotional) nature of info on LEK from participants.

In other words, the difficulty here concerns how subjective info "is valued. The interviewees often regard LEK as subjective (associated with opinions and feelings), which creates challenges in how to use the knowledge....LEK can be very emotional, because ordinary residents may not be able to argue clearly about their experiences and opinions. However, as the interviewed planners note, emotional stances have their reasons, and a good planner tries to interpret what is at the root of these emotional outbursts. Furthermore, local participants can appreciate different elements of nature in different ways. For instance, for some residents certain trees can be close and familiar, i.e., associated with numerous memories, whereas others may see the same trees as 'just shading.' A planner, however, has to take both into consideration as a part of the planning process. The different perspective of each resident may thus become a dilemma for the planner who tries to incorporate these conflicting messages." (J. Kohl 2005)

*A Critical Assessment*

The lack of general agreement on the difficulties of how to collect and use local ecological knowledge (LEK) in the process of implementing sustainable urban planning raises the serious issue concerning the extent of possibility and desirablity of sustainability in urban planning.

After all, to what extent is sustainability possible? In other words, if collecting good information is difficult to obtain, then how exactly can sustainability be successfully implemented in urban planning?

In addition, to what extent is sustainability desirable? In other words, if not all participants are involved, and if their opinions are not necessarily to be included, depending on which model is to be accepted—then how can sustainability be valued by all in society?

Therefore, it is easier said than done, when the issue of sustainability in urban planning is raised, as shown in the summary in the previous sub-section.

Yet, the summary can also teach us something relevant to the ontolological principles in existential dialectics, and good examples include the absoluteness-relativeness principle, the partiality-totality principle, the predictability-unpredictability principle, the denseness-emptiness principle, the preciseness-vagueness principle, the simpleness-complicatedness principle, the order-chaos principle, the evolution-transformtion principle, the symmetry-asymmetry principle, and the regression-progression principle.

For instance, in relation to the absoluteness-relativeness principle, if there is absolute analysis (e.g., the sole concern of greedy individuals with using natural resources for their own needs), there is relative concern too (e.g., the dual concerns with using and exploiting natural resources in the relative viewpoint of balancing with nature, so that what is valid from one standpoint may not be so from another).

In relation to the partiality-totality principle, if there is partial analysis (e.g., the narrow focus by each individual for his own needs in using natural resources), there is the totalistic analysis too (e.g., the bigger picture of ecolological interdependence connecting all individuals in their relation to nature).

In relation to the predictability-unpredictability principle, if there is predictability (e.g., the concern with sustainability when the tipping point of violating the balance with nature is crossed), there is unpredictability too (e.g.,. the difficulty to know in advance which specific ways to collect and use LEK from the participants in a given community).

In relation to the denseness-emptiness principle, if there is denseness (e.g., the greater use of resources as the number of population in a habitat increases), there is also emptiness (e.g., the need to restore the balance with nature by making a habitat less dense, say, by way of "low-intensity habitation").

In relation to the preciseness-vagueness principle, if there is vagueness (e.g., the conceptualization of only three general models of LEK in sustainabile urban planning), there is also the need for preciseness (e.g., the further subdivision of M2a and M2b in the context of LEK).

In relation to the simpleness-complicatedness principle, if there is simplistic analysis (e.g., the call for sustainable urban planning with the use of LEK), there is the complicated challenge (e.g., the complicated analysis of different serious difficulties in collecting and using LEK).

In relation to the order-chaos principle, if there is order (e.g., the patterns of ecological interdependence in nature), there is chaos too (e.g., the uncertain outcomes in a complex system).

In relation to the evolution-transformation principle, if there is evolution (e.g., the natural evolution of humans to live in relation to nature), there is also transformation (e.g., the technological transformation of urban space to balance with nature for sustainable urban planning).

In relation to the symmetry-asymmetry principle, if there is symmetry (e.g., the co-existence of different participants with their diverse opinions and emotions on LEK), there is asymmetry (e.g., the use of LEK from some of the participants and the rejection of LEK from others ).

And in relation to the regression-progression principle, if there is progress (e.g., the better understanding of ecological interdependence over time), there is regression (e.g., the lack of general agreement on collecting and using LEK, with the subsequent serious questioning on the extent of possibility and desirability of sustainability in urban planning).

# Density and Society

An even more complicated way to examine the issue of density in relation to sustainability in urban planning is from the perspective of society.

Consider a few case studies below, that is, in relation to (2.4.1) urban pollution and institutional failures, (2.4.2) urban renewal and structure injustice, and (2.4.3) sustainable technology and system complexity—to be discussed hereafter, in that order.

## Urban Pollution and Institutional Failures

The problem of pollution in cities is something that urban planners need to pay attention to, because of its damaging consequences to nature, society, and the mind, for instance. But the fight against urban pollution can still fail, especially though not exclusively when contextualized in the domain of social institutions.

To start, pollution is a form of "negative externalities," since the latter can also include "unsafe auto accidents, workplace injuries, product hazards such as lead paint, etc. But pollution probably has the longest and most tragic tradition." (S. Kangas 2008)

For instance, "air pollution alone kills 64,000 people in the U.S. each year. Dioxins are estimated to kill another 26,000 to 260,000. And this isn't even counting the environmental devastation and loss of life caused by water contamination, radiation, acid rain, global warming, ozone depletion, etc. The final U.S. mortality rate may reach the hundreds of thousands each year. And this is only the fatalities—nonfatal environmental diseases are even worse." (S. Kangas 2008; USEPA 1994; NRDC 1996)

Conventionally, there are two approaches to solving urban pollution, namely, (a) market solution and (b) government solution. Let's consider them one by one, and see if they are successful, in what follows.

### *Market Solution*

A good illustration of the market solution to the problem of urban pollution is the original proposal by Ronald Coase, or what is now known as "the Coase theorem."

The Coase theorem thus states: (S. Kangas 2008)

- "Given well-defined property rights, low bargaining costs, perfect competition, perfect information and the absence of wealth and income effects, resources will be used efficiently and identically regardless of who owns them."

In other words, there are some basic assumptions in the Coase theorem to be examined, in relation to (i) property rights, (ii) efficient and identical use of resources, (iii) transaction costs in bargaining, (iv) perfect competition and information, and (v) wealth and income effects.

But are these assumptions valid? Steve Kangas (2008) answered in the negative. Why?

(i) Firstly, in relation to the issue about property rights—Coase argued that "many disputes over resources stem from the fact that no one owns them. Or—nearly as bad—everyone owns them, as in the case of public property. However, these disputes could be resolved if the unclaimed resources were divided up as private property. Now if someone wants to use your property, you could charge them a fee. Or if they abuse your property, you can sue them in court. Assigning property rights greatly enhances the ability to resolve disputes over the use and abuse of resources." (S. Kangas 2008)

But the problem here is that, as Kangas (2008) wrote, "property rights are actually a bundle of many different rights. For example, you have a right to prevent airplanes from flying ten feet above your property, but you do not have a right to prevent them from flying four miles overhead. Likewise, you may have a right to prevent a factory next door from choking your air with carbon dioxide. But humans also exhale carbon dioxide as a natural by-product of their metabolism. Thus, we cannot establish a single, monolithic 'property right' that prevents people from affecting your property in any way, shape or form. In defining property rights, we must identify the bundle of rights most appropriate for the owner. This is important because Coase argues that we should be able to sell individual 'sticks' from this 'bundle'—for example, selling a factory the right to pollute your air."

In addition, "property rights are irrelevant to the environmental problem. The population explosion is a good example. Scientists attribute most of our environmental problems to the world's burgeoning population, which places ever greater demands on the environment's limited resources….Strengthening property rights would hardly affect this trend." (S. Kangas 2008)

(ii) Secondly, in relation to the efficient and identical use of resources—Coase argued something like this: "Suppose that a chemical factory is dumping pollution into a lake. But there is a group of fisherman on the lake who earn their living by catching fish, and the pollution greatly reduces their yearly catch." (S. Kangas 2008)

So, what is the solution? Well, it all depends on the amount of pollution damage (for compensation) and the amount of a pollution-cleaning equiment. If the amount of pollution damage is more than the amount of a pollution-cleaning equipment, it is better to buy a pollution-cleaning equipment, either by the factory (if the fishermen own the lake) or by the fisherman (if the factory owns the lake).

But the problem here is, What if the amount of pollution damage is less than the amount of a pollution-cleaning equipment? Well, in this case, it is better to keep polluting by the factory, and if the fishermen own the lake, the factory can simply pay them a lesser amount in return.

So, the bottom line is, as Kangas (2008) argued, "the Coase theorem doesn't completely eliminate pollution even in principle. Defenders of the Coase theorem point out that complete elimination is probably impossible anyway."

(iii) Thirdly, in relation to transaction costs in bargaining—Coase claimed that transaction costs can be reduced in free market. But how exactly?

The essential problem here is that "[t]ransaction costs are the costs of negotiating and agreement. If there are numerous parties involved, or bitter legal disputes with teams of legal eagles, then the costs of coming to an agreement may exceed its benefits." (S. Kangas 2008)

After all, there are numerous forms of "transaction costs," and good examples include "the expense involved in bringing many parties to an agreement," "search and information costs in finding the parties involved and informing them of exchange opportunities," "enclosure costs" (viz., "fencing the property"), "monitoring and policing costs" (viz., "to ensure that agreements are kept"), and "prosecution costs for trespassers and violators." (S. Kangas 2008)

And all these transaction costs are further compounded by another problem, in that, when the parties in question can be too many (e.g., millions of residents in a big city) in "the case of pollution, this is an insurmountable problem. It is impossible for millions of commuters and residents in [a metropolis like] Los Angeles to reach a unanimous agreement on controlling smog." (S. Kangas 2008)

In the end, "how can transaction costs be reduced? Ultimately, Coase admits they can't. But a recurring theme in his writings is that governments are usually in no position to do better." (S. Kangas 2008)

(iv) Fourthly, in relation to perfect competition and information—Coase argued that they are important for free market to function.

Yet, the inherent problem here is that "perfect competition and information" do not quite exist in the real world.

For instance, in relation to free competition, "the market features a large number of competitors, homogeneous goods, free entry and exit, and perfect information. This results in a high degree of efficiency: namely, lower prices and higher quality of goods." (S. Kangas 2008)

But perfect competition often fails, and imperfect forms of competition take over, as in the case of "monopolies or oligopolies….These institutions are notorious for raising their prices, lowering the quality of their goods, exploiting customers, and otherwise engaging in inefficient behavior." (S. Kangas 2008)

And in relation to perfect information, it also does not exist in real life, but without it, however, "the incentive of firms to become even more efficient will not be as great. Also, perfect information is required if firms are to be aware of bargaining possibilities with others. This includes knowing the financial state of one's competitors and bargainers. Without such knowledge, externalities will persist, unrecognized and unresolved." (S. Kangas 2008)

In fact, the problem of perfect information is further worsened when there are differences in barganing power. As Kangas (2008) explained, "[a]n extreme example of this was the white man's settlement of the American West. When Native Americans attempted to use the land as common property, gold miners who had staked it as private property appealed to the U.S. Cavalry to force a better 'agreement.' In modern times, lawyers, not guns, constitute bargaining power. A giant corporation with its team of legal eagles, researchers, encyclopedic information and business and political connections will be able to win the better deal from an average Joe with no resources. Coasians also tend to ignore one of the most basic of bargaining tools: the threat. Inefficient agreements could very well result when one party has an advantage over another."

And (v) in relation to the absence of wealth and income effects—Coase argued that "apart from such cataclysmic events as the abolition of slavery, these effects will normally be so insignificant that they can safely be neglected." (S. Kangas 2008; R. Coase 1988)

But is this really true? Steve Kangas begged to differ. By definition, wealth and income effects to "the change of wealth or income that occurs when you are awarded property or property rights," and this is important because "a different assignment of property rights changes society's overall supply and demand." (S. Kangas 2008)

For instance, in the previous factory-and-fishermen analysis, "[i]f the factory has to buy the equipment, then the price of their widgets goes up. If the fishermen have to buy it, then the price of fish goes up. These differences will reverberate through the entire economy." (S. Kangas 2008)

In addition, "[t]here is also a second type of income effect: most people desire a higher price for selling a right than buying a right. Suppose someone is dumping garbage in your yard. The odds are high that you would charge them much more to continue dumping than you would pay them to stop it. So the income that either party makes from this relationship will differ, depending on who gets the property right." (S. Kangas 2008)

So, with the problems facing these five assumptions in mind, the final outcome to fight urban pollution with the market approach is far from certain.

*Government Solution*

But if there are market failures in solving urban pollution, how about government solution?

Traditionally, there are two major ways to give the government a role to solve the urban pollution problem, namey, (i) the "polluter pays" solution and (ii) the "Pigouvian tax" solution.

(i) In relation to the "polluter pays" solution, "polluters are responsible for cleaning up their pollution. A major selling point of 'polluter pays' is its relative efficiency. Scientists need to conduct only a few studies to determine, say, how much mercury can be tolerated in the bloodstream before becoming a health hazard, and thus what level of environmental mercury is acceptable. Government then establishes these levels as society-wide norms, and needs only to inspect a random sampling of businesses to encourage compliance. If polluters exceed these norms, they must pay the cleanup costs. This might raise the price of their products, but it 'internalizes' the externality, by passing the full cost of the product on to the consumer. This may raise the prices of many products, shift supply and demand, and perhaps even bankrupt some companies while creating new ones." (S. Kangas 2008)

And (ii), in relation to the "Pigouvian tax" solution (named after Arthur Pigou)—the government taxes a company for each unit of pollution that it emits. Companies therefore have an incentive to reduce pollution as much as possible. Presumably the extra cost gets passed on to the consumer, and the government uses the funds to fight pollution. But this is much less efficient and direct than 'polluter pays.'" (S. Kangas 2008)

In either case, however, is government intervention really effective to fight urban pollution? For some critics, the answer is far from satisfactory, and a good argument is provided by those scholars in Public Choice Theory, who are quick to point out that the polluters can get away from the taxes and norms, because of the tendency of "rent seeking" by "interest groups to lobby for laws and regulations that provide them a guaranteed benefit (rent)." (WK 2007a; J. Shaw 2002)

There are three main types of "rent-seekers" here, namely, "legislators, administrators, and regulated entities." (WK 2007a) The relationships among the three groups, though complicated, are to ensure their own interests, often at the expense of the public (like the ordinary voters).

For instance, "legislators will try to change laws to ensure their continued incumbency, such as gerrymandering," and "administrators (bureaucrats) may seek to advance their power and budget authority," while "regulated entities may seek to obtain barriers to entry or other special provisions of the law that reduce competition, increase subsidization," or something else. (WK 2007a)

The public (like the ordinary voters), on the other hand, are often uninformed and uninterested in monitoring "government effectively. Anthony Downs [1957], in one of the earliest public choice books, *An Economic Theory of Democracy*, pointed out that the voter is largely ignorant of political issues....Even though the result of an election may be very important, an individual's vote rarely decides an election. Thus, the direct impact of casting a well-informed vote is almost nil; the voter has virtually no chance to determine the outcome of the election. So spending time following the issues is not personally worthwhile for the voter. Evidence for this claim is found in the fact that public opinion polls consistently find that less than half of all voting-age Americans can name their own congressional representative." (J. Shaw 2002)

Because of this weak monitoring by the public, legislators, although they "are expected to pursue the 'public interest,'" often "make decisions on how to use other people's resources, not their own....There is no direct reward for fighting powerful interest groups in order to confer benefits on a public that is not even aware of the benefits or of who conferred them....In contrast, interest groups are organized by people with very strong gains to be made from governmental action. They provide politicians with campaign funds and campaign workers. In return they receive at least the 'ear' of the politician and often gain support for their goals." (J. Shaw 2002)

In addition, "many regulatory agencies appear to be 'captured by special interests....The 'capture' theory was introduced by the late George Stigler, a Nobel Laureate....Capture occurs because bureaucrats do not have a profit goal to guide their behavior. Instead, they usually are in government because they have a goal or mission. They rely on Congress for their budgets, and often the people who will benefit from their mission can influence Congress to provide more funds. Thus interest groups—who may be as diverse as lobbyists for regulated industries or leaders of...[other] groups—become important to them. Such interrelationships can lead to bureaucrats being captured by interest groups." (J. Shaw 2002)

*A Lesson to Learn*

The lesson to learn from this debate is that neither the free market nor the government, at this stage of human history, can ensure the provision of a pollution-free urban environment, which is used as a case study of sustainability in urban planning here.

The reasons are that there exist both market and government failures, which then make it very difficult for urban planners to design a metropolis which can sustain for the long-term.

But then, the debate can also reveal something refreshing in relation to the ontological principles in existential dialectics, and good examples include the absoluteness-relativeness principle, the partiality-totality principle, the denseness-emptiness principle, the simpleness-complicatedness principle, the openness-hiddenness principle, the finiteness-transfiniteness principle, the convention-novelty principle, the symmetry-asymmetry principle, the regression-progression principle, and the same-difference principe.

For instance, in relation to the absoluteness-relativeness principle, if there is absolute analysis (e.g., Coase's obsession with the free market to solve the pollution problem), there is relative counterpart (e.g., the more balanced analysis of the pros and cons of both the market and government solutions, so that what is valid from one standpoint may not be so from another).

In relation to the partiality-totality principle, if there is partial viewpoint (e.g., the individual focus on his own self-interest in deciding whether to pollute or not), there is likewise totalistic concern (e.g., the tragedy of the commons when viewed from the larger picture).

In relation to the denseness-emptiness principle, if there is denseness (e.g., the occurrence of pollution, be it noise, air, or something else, in a densely populated environment), there is also emptiness (e.g., the search for an escape from urban density, such as fighting pollution or seeking a greener open space).

In relation to the simpleness-complicatedness principle, if there is simplistic analysis (e.g., Coase's naïve faith on the free market to fight pollution), there is also complicated counterpart (e.g., a more complicated analysis of multiple actors and factors in the process of fighting pollution, as shown in the market and government models).

In relation to the openness-hiddenness principle, if there is openness (e.g., the exploration of the serious problem of pollution for the public), there is hiddenness (e.g., the hidden interests of politicians, regulators, and interest groups, in light of the weak monitoring by the public).

In relation to the finiteness-transfiniteness principle, if there is finiteness (e.g., the multipe actors in the collusion among regulators, legislators, and interest groups), there is also transfiniteness (e.g., the millions, or more than millions, of voters or residents in a metropolis who are weak in monitoring government effectiveness).

In relation to the convention-novelty principle, if there is conventional wisdom (e.g, the traditional "polluter pays" solution), there is likewise novel challenge (e.g., the original Coase's theorem to fight pollution).

In relation to the symmetry-asymmetry principle, if there is symmetry (e.g., the co-existence of different approaches to fight urban pollution), there is asymmetry (e.g., the popularity of the Coase's theorem among free marketers, and the acceptance of government intervention among those on the Left).

In relation to the regression-progression principle, if there is progress (e.g., the better understanding of the seriousness of urban pollution and the available means to counter it), there is regression too (e.g., the collusion among politicians, regulators, and interest groups, as shown in public choice theory).

And in relation to the same-difference principle, if there is difference in outcome (e.g., the lessening of urban pollution over time in some cases), there is similarity in outcome too (e.g., the persistent existence of collusion among different groups with their ulterior interests).

## Urban Renewal and Structural Injustice

Another way to explore the possibility and desirability of sustainability with urban density as a case study concerns the phenomenon of "urban renewal," especially though not exclusively in the context of social structure. (WK 2008p) So, a good question here is, Does urban renewal really work?

*Defining "Urban Renewal"*

The words "urban renewal" (also known as "urban regeneration") refers to "a controversial U.S. program of land re-development in areas of moderate to high density urban land use. This process began as an intense phase in the late 1940s and continued into the late 1970s with traces of it still occurring in the early 1980s." (WK 2008p)

Urban renewal, as an urban planning vision, has "a major impact…in the history and demographics of cities around the world, including: Beijing, China, Melbourne, Victoria; Saint John, New Brunswick; Glasgow, Scotland; Boston, Massachusetts; Warsaw, Poland; San Francisco, California; and Bilbao, Spain. Commonly cited examples include Canary Wharf, in London, and Cardiff in Wales." (WK 2008p)

*A Brief History*

In the United States, urban renewal "goes back to the work of Robert Moses in the redevelopment of New York City and New York State from the 1930s into the 1970s. Moses directed the construction of new bridges, highways, housing projects, and public parks….Moses is responsible for most major traffic arteries in the city and for its largest parks, except Central Park and Prospect Park." (WK 2008p)

Shortly later, in 1934, "the National Housing Act of 1934…established the Federal Housing Administration (FHA) to improve housing conditions and standards, and later led to the formation of the Department of Housing and Urban Development (HUD). While it was designed to develop housing for poor residents of urban areas, that act also required cities to target specific areas and neighborhoods for different racial groups, and certain areas of cities were not eligible to receive loans at all. This meant that ethnic minorities could only secure mortgages in certain areas, and resulted in a large increase in the residential racial segregation in the United States." (WK 2008p)

In other cases, public housing projects were implemented, but they "required that slum housing be demolished prior to any construction." (WK 2008p) A good example involves Pttsburgh in 1950, where "a large section of downtown at the heart of the city was demolished, converted to parks, office buildings, and a sports arena and renamed the Golden Triangle….Other neighborhoods were also subjected to urban renewal, but with mixed results. Some areas did improve, while other areas, such as East Liberty and Lower Hill declined following ambitious projects that shifted traffic patterns, blocked streets to vehicular traffic, isolated or divided neighborhoods with highways, and removed large numbers of ethnic and minority residents." (WK 2008p)

In 1956, "the Federal-Aid Highway Act gave state and federal government complete control over new highways, and often they were routed directly through vibrant urban neighborhoods—isolating or destroying many—since the focus of the program was to bring traffic in and out of the central cores of cities as expeditiously as possible and nine out of every ten dollars spent came from the federal government. This resulted in a serious degradation of the tax bases of many cities, isolated entire neighborhoods, and meant that existing commercial districts were bypassed by the majority of commuters. Segregation continued to increase as communities were displaced and many African Americans and Latinos were left with no other option than moving into public housing while Whites moved to the suburbs in ever-greater numbers." (WK 2008p)

And Boston is another example of urban renewal, with both promises and failures, since "almost a third of the old city was demolished—including the historic West End—to make way for a new highway, low- and moderate-income high-rises (which eventually became luxury housing), and new government and commercial buildings. Later, this would be seen a tragedy by many residents and urban planners, and one of the centerpieces of the redevelopment—Government Center—is still considered an example of the excesses of urban renewal....[And] the transformation from the Columbia Point housing projects, opened in 1954,...had been later left to deteriorate significantly, to be totally re-vitalised starting in 1984 by the development team at Corcoran, Mullins, Jennison and later renamed the Harbor Point Apartments." (WK 2008p; B. Kamin 1997; J. Roessner 2000)

*Urban Renewal from a Critical Perspective*

The success of urban renewal in the U.S. experience (as briefly summarized above) cannot be denied. Yet, at what costs?

For instance, Jane Jacobs in *The Death and Life of Great American Cities* (1961) is "one of the first—and strongest—critiques of contemporary large-scale urban renewal. However, it would still be a few years before organized movements began to oppose urban renewal." (WK 2008p)

As an illustration, in 1964, "the Civil Rights Act removed racial deed restrictions on housing. This began desegregation of residential neighborhoods, but redlining continued to mean that real estate agents continued to steer ethnic minorities to certain areas. The riots that swept cities across the country from 1965 to 1967 damaged or destroyed additional areas of major cities—most drastically in Detroit during the 12th Street Riot." (WK 2008p)

By the early 1970's, "many major cities developed opposition to the sweeping urban-renewal plans for their cities. In Boston, community activists halted construction of the proposed Southwest Expressway—but only after a three-mile long stretch of land had been cleared. In San Francisco, Joseph Alioto was the first mayor to publicly repudiate the policy of urban renewal, and with the backing of community groups, forced the state to end construction of highways through the heart of the city. Between 1956 and 1966, more than 12% of the people in Atlanta lost their homes to urban renewal, expressways, and a downtown building boom turned the city into the showcase of the New South in the 1970's and 1980's." (WK 2008p)

By then, "renewal often resulted in the creation of urban sprawl and vast areas of cities being demolished and replaced by freeways and expressways, housing projects, and vacant lots, many of which still remain vacant at the beginning of the 21st century."(WK 2008p; J. Lobbia 1999)

In 1968, the policy of "urban renewal" was questioned to the point that "community development" became a new version of urban renewal, as shown in "the Housing and Community Development Act of 1974," which "established the Community Development Block Grant program (CDBG) which began in earnest the focus on redevelopment of existing neighborhoods and properties, rather than demolition of substandard housing and economically depressed areas." (WK 2008p)

Community development, in this sense, is "a process of urban renewal which enables local citizens to have greater control and ownership of the direction of their community and the way in which it overcomes market failure. This supports important themes in urban renewal today, such as participation,sustainability and trust—and government acting as advocate and 'enabler,' rather than an instrument of command and control." (WK 2008p)

In the end, "urban renewal never lived up to the hopes of its original proponents….It has been hotly debated by politicians, urban planners, civic leaders, and current and former residents of the areas where urban renewal took place in the 1950s, 1960s, and 1970s. It has brought economic and cultural development to many cities, but often at a great cost to low-income and minority communities living in them. It has also played a role in the economic devastation faced by many of the major industrial cities in the United States since the 1940s. Urban renewal continues to evolve as successes and failures are examined and new models of development and redevelopment are tested and implemented." (WK 2008p)

In this light, it is no wonder for James Baldwin to "famously dubb…Urban Renewal 'Negro Removal,'" and urban renewal "is seen…by opponents as a regressive mechanism for enriching the wealthy at the expense of taxpayers and the poor. It carries a high cost to existing communities, and in many cases resulted in the destruction of vibrant—if run-down —neighborhoods." (WK 2008p; D. Fitzpatrick 2000; W. Domhoff 1983; MSNBC 2005)

This is true, even when "urban renewal evolved into a policy based less on destruction and more on renovation and investment, and today is an integral part of many local governments, often combined with small and big business incentives. But even in this adapted form, Urban Renewal projects are still widely accused of abuse and corruption." (WK 2008p)

*The Usefulness of the Debate*

This critical awakening on the once popular policy of urban renewal can reveal something illuminating in relation to existential dialectics, and good examples include the absoluteness-relativeness principle, the denseness-emptiness principle, the simpleness-complicatedness principle, the expansion-contraction principle, the convention-novelty principle, the symmetry-asymmetry principle, and the regression-progression principle.

For instance, in relation to the absoluteness-relativeness principle, if there is absolute viewpoint (e.g., the euphoric advocation of urban renewal, especially by Robert Moses in 1930's), there is also relative counterpart (e.g., the more balanced analysis of both benefits for the rich and the costs for the poor and the minorities in urban renewal, so that what is valid from one standpoint may not be so from another).

In relation to the denseness-emptiness principle, if there is denseness (e.g, the policy of urban renewal to cope with the problem of urban density), there is also emptiness (e.g., the subsequent urge for less dense environment by way of urban sprawl and later suburban sprawl, in the aftermath of urban renewal).

In relation to the simpleness-complicatedness principle, if there is simplistic analysis (e.g., the single-minded focus on restructuring a vast section of a city for commercial, government, and social services in traditional urban planning), there is likewise its complicated counterpart (e.g., the more complicated consideration of the plight of the ecological environment, the social structure, and community development in any urban planning).

In relation to the expansion-contraction principle, if there is expansion (e.g., the occurrence of urban sprawl), there is likewise its contraction (e.g., the problem of urban decay over time).

In relation to the convention-novelty principle, if there is conventional wisdom (e.g., the old-fashioned advocation of urban renewal), there is novel challenge (e.g., the original replacement of "urban renewal" with "community development").

In relation to the symmetry-asymmetry principle, if there is symmetry (e.g., the co-existence of both the rich and the poor in urban areas), there is asymmetry (e.g., nice neighborhoods for the rich and bad ones for the poor).

And in relation to the regression-progression principle, if there is progress (e.g., the awareness of the need for urban renewal with the consideration of the plight of minorities and the ecological environment), there is its regression (e.g., the persistent gap between the rich and the poor in urban habitation, and the serious recurrence of ecological problem).

## Sustainable Technology and System Complexity

There is still another way to explore the possibility and desirability of sustainability in urban density, this time, in relation to social systems, or more precisely, in the context of how sustainable technology can still fail, in a complex social system.

### *The "Mousetrap" Theory of Sustainable Technology*

The work of John Sterman (2008) is revealing enough here, since he debunked "some popular proposals for addressing climate change, with sobering case studies that demonstrate why 'technological solutions are not enough to address the problem of creating a sustainable world.'"

The fashionable talk for a quick fix of the problem of "climate change" (in what is now known as the "moursetrap" theory of sustainable technology) can be summarized like this: "New technology…will drive down the cost of renewable energy, increase demand for carbon-free renewables and displace fossil fuels. New energy markets emerge, after a regulatory nudge or two from the government, or some incentives and emissions fees." (J. Sterman 2008)

In other words, an increase of technological innovation will eventually lead to lower cost of renewable energy, with the help of government subsidies at the beginning, which in turn will yield lower price of renewable energy, higher demand for renewable energy, and higher price of (conventional) fossil fuels (to discourage the use of conventional gasoline-fuel vehicles).

But this theory is simplistic enough, for Sterman, since highly innovative technologies do not guarantee high social use and adoption, and, conversely, low innovative technologies do not necessarily yield low social use and adoption.

For instance, "the Sony Betamax video recorder" has high quality of technological innovation but low social use and adoption; in fact, the market for Sony Betamax did not make it in the end. (J. Sterman 2008)

On the other hand, the practice of astrology has low technological innovation but has high social use and adoption, and the same applies to "qwerty keyborad." (J. Sterman 2008).

The essential point here, for Sterman (2008), is that, more often than not, "opportunities that would significantly reduce our carbon footprint yet have been ignored by society at large, such as improving fuel efficiency, and insulating buildings. Our rejection of these opportunities suggests we can't comprehend 'the complexity of systems in which we are embedded and into which we deploy technologies,' particularly the concept of feedback."

*The Challenge of System Complexity*

In a thought experiment, Sterman (2008) analyzed "the introduction of a hydrogen-based, zero tailpipe emission alternative fuel vehicle (AFV) into California—a conceivable leap toward creating an ecologically and economically sustainable transportation system. The government kick-starts the AFV market, rolling out fuel stations in urban centers, and essentially subsidizing the transition for a decade. You'd expect this AFV eventually to command at least 50% of the market share. But when Sterman runs his simulations, the AFV stagnates at around 25%."

Why? The answer is that, sustainable technology by itself cannot do the work, since, in the current example, "if fuel stations are not distributed through even the remotest parts of the state, people worry about where they'll find fuel, leading to weak demand for AFVs. This is 'only one of the many reinforcing feedbacks which create strong barriers to the entry of technologies which are as good or better than incumbent technologies'....Even an AFV with higher fuel efficiency can't win market share." (J. Sterman 2008)

In fact, Sterman (2008) showed that the introduction of alternative fuel vehicles (AFVs) has failed in many cases around the world, be they about "compressed natural gas," "diesel," or "ethanol."

But "the models offer some faint promise. When Sterman puts more fuel stations in rural areas, the AFV market succeeds," but only after a very long period of time—assuming also, of course, that everything else goes smoothly.

*Some Caveats*

For the critics, however, Sterman's model is not complex enough, as he himself admitted. Here are two good examples.

Firstly, there are "large uncertainties in technical, economic and other parameters" which need to be considered for the study of system complexity, but his models do not focus on them.

And secondly, there are also "unintended policy impacts and counterintuitive dynamics" which are important. For instance, any introduction of AFVs can also have unintended outcomes like "surplused used conventional vehicles" which then "depress car prices, delay AFV adoption," and even increase "fuel tax" to "maintain constant revenue" after the fall of "excise tax reveunes" when "gasoline consumption falls" (with the introduction of AFV's). (J. Sterman 2008)

Yet, Sterman's point here is that any quick fix of the problem of climate change with the use of sustainable technology is easier said than done. After all, there is already "a century of learning" of how to "dethrone…gasoline" by "cost reduction" and the challenge to overcome any "lock-in to existing infrastructure," but the outcome has been far from satisfactory. (J.Sterman 2008)

In fact, to "create…an economically sustainable AFV market requires aggressive, long-lived policies to drive consumer awareness and adoption, vehicle costs and performance, and fuel infrastructure over the tipping point," often with things getting much worse before getting better in the long run. (J. Sterman 2008)

In the end, "creating a sustainable world…depends on complex feedbacks among technology, economics, beliefs and behavior, social forces, politics, culture," or, in short, everything in a complex social system. (J. Sterman 2008) Not only "good technologies do not necessarily win," but also there is "no purely technical solution" to create a systainable world.

In fact, this is something that I already went to great lengths to explain in Chapter Two of *The Future of Human Civilization* (*FHC*), Chapter Seven of *The Future of Capitalism and Democracy* (*FCD*) on the illusion of "e-topia," and, for that matter, *The Future of Post-Human Engineering* (*FPHE*).

Be that as it may, the debate here shows us something about the ontological principles in existential dialectics, and some examples are the absoluteness-relativeness principle, the partiality-totality principle, the predictability-unpredictability principle, the denseness-emptiness principle, the simpleness-complicatedness principle, the theory-praxis principle, the convention-novelty principle, the evolution-transformation principle, the symmetry-asymmetry principle, and the regression-progression principle.

For instance, in relation to the absoluteness-relativeness principle, if there is absolute analysis (e.g., the one-dimensional focus on the building of sustainable technologies to fight the problem of climate change), there is also relative analysis (e.g., the more balanced linkage between the invention of technologies and the wider social and cultural factors, so that what is valid from one standpoint may not be so from another).

In relation to the partiality-totality principle, if there is partial analysis (e.g, the focus on the individual factors for someone to decide whether or not to adopt an AFV), there is also totalistic analysis (e.g., the concern with the larger, uncertain outcomes of the system as a whole in the AFV market)..

In relation to the predictability-unpredictability principle, if there is predictability (e.g., the prediction of more availability of AFVs over time), there is unpredictability (e.g., the inability to tell in advance which AFV markets will succeed and which ones will not).

In relation to the denseness-emptiness principle, if there is denseness (e.g., the highly croweded urban environment, with its ever increasing demand for energy resources, like gasoline consumption), there is emptiness too (e.g., the search for smaller AFVs and greener space, to minimize urban crowededness).

In relation to the simpleness-complicatedness principle, if there is simplistic analysis (e.g., the advocation of AFVs by its proponents, with little detailed analysis of the wider challenge of sustainability), there is also complicated analysis too (e.g., the more complicated account for AFVs, in relation to the multiple possibilities between the quality of technological innovation and its use or adoption).

In relation to the theory-praxis principle, if there is theoretical construction (e.g., the mousetrap theory of sustainable technologies), there is also its practical application (e.g., the practical manufacturing of AFVs by different car firms over time, be they successful or not).

In relation to the convention-novelty principle, if there is conventional wisdom (e.g., the popular talk of AFVs to fight the problem of climate change), there is novel challenge (e.g., the original view by Sterman on how bad things can happen to sustainable technologies).

In relation to the evolution-transformation principle, if there is evolution (e.g., the natural evolution of humans to walk uprght and run), there is also transformation (e.g., the technological construction of AFVs for humans to move around faster and more cheaply).

In relation to the symmetry-asymmetry principle, if there is symmetry (e.g., the co-existence of different vehicles, both gasoline-fuel and alternative-fuel), there is asymmetry (e.g., the popularity of gasoline-fuel vehicles in the older days, and the increasing preference for AFVs in our time).

And in relation to the regression-progression principle, if there is progress (e.g., the invention of AFVs), there is regression too (e.g., the failure of many AFV markets in the last century, and its unintended policy impacts).

# Density and Culture

The last part of the chapter concerns the last perspective to be considered, this time, in relation to culture, in the context of sustainability with urban density as a case study.

Consider a few illustrations in what follows, namely, (2.5.1) urban planning and the religious search for order and purity, and (2.5.2) aesthetics and soundscape planning—in that order.

## Urban Planning and the Religious Search for Order and Purity

Urban planning has long had a religious connection. For example, Lewis Mumford (1967) in *The Myth of the Machine: Technics and Human Development* argued that "historians have over-emphasized the role of the machine and technique in humanity's development. Even before language was available, our ancestors 'had a dim consciousness of the mystery of their own being'....Planning is not so much a tool we use to shape our urban environments as it is an expression of a continued belief in myth, magic and religion." (D. Engwicht 2009) When contextualized within urban planning, the expression of the religious belief is often revealed in the search for "order."

Here are some examples. In ancient Egypt, the city of Tel-el-Amarna "is laid out in grid fashion and is dated around 2670 B.C. It is cited by many as the first example of a planned city....What is significant is that religious beliefs motivated the establishment of a whole new town—a significant act of planning." (D. Engwicht 2009)

In China, there has been the tradition of "Feng Shui," which is "to study and expound the shapes which the spiritual forces of nature have produced and to prescribe the ways in which all buildings, roads, bridges, canals and railways must conform to them." (D. Engwicht 2009; P. Hall 1988) Of course, there can be practical purposes for this ordering, like "improving health (sewerage systems) or for safety (defense walls and moats)," but "much of it was tied to the religious impulse—an attempt to grasp the meaning of life and the universe." (D. Engwicht 2009)

*The Role of Christianity*

And in Christian Europe, "the basis of the Christian mythology is the Bible, which starts with a cosmos which is dark, empty, void and without form—in a state of chaos. God broods over this chaos, producing paradise which comes in the form of a garden. Adam and Eve succumb to evil, introducing suffering, work and a certain kind of disorder into the created realm. They bear two children. The elder, Cain, a tiller of the soil, murders his younger brother and is then banished to become a refugee and wanderer. However, he settles in the land of Nod and builds a city. Thus the first city is viewed by the Christian theologian as being conceived in rebellion and violence." (D. Engwicht 2009)

The idea of the "evil city" (as opposed to the "holy city" or the "city of God"), further took shape in the city of "Babel (Babylon), a symbol for Christians of humanity's arrogance and the construct of the city as an act of idolatry." (D. Engwicht 2009)

Bill Kellermann (1989) thus explained on the symbol of Babel as the evil city for the Christians: "The city, then, is not a Promethean act reaching up to God. It is the act of making an identity by making a world, an urban environment, a great city. It is the act of excluding God from creation. It is incipient and express idolatry." Later, this act of idolatry eventually led to moral decadence and sexual deviance in the city.

Later, "during the reign of Nebuchadnezzar, Babylon became one of the greatest cities of antiquity. In 597 Nebuchadnezzar conquered Jerusalem and took the Jews into exile in Babylon. Some of the writings generated while in exile (Isaiah, Jeremiah, Ezekiel, Daniel) became part of Jewish scripture and their theme is an outpouring of hate and condemnation against the evil city." (D. Engwicht 2009)

This religious condemnation of the evil city is not trivial, since, in the case of Christian Europe, "this imagery of the city as the centre of evil, particularly sexual sin, has exercised a powerful influence for at least the last 2,500 years—ever since the Babylonian exile," and, as Hall (1988) thus observed, "most of the philosophical founders of the planning movement continued to be obsessed with the evils of the congested Victorian slum city," especially in the context of "a strong preoccupation with sexual deviance." (D. Engwicht 2009)

By contrast, the "holy city" (or the "city of God") is just the opposite of the "evil city," since the former "sees the city as God's home, the Celestial City, instrument of redemption, place of wholeness and justice. For example, God instructed Moses to build cities of refuge where those who were refugees could come. And throughout the Bible Jerusalem, is constantly referred to as a holy city or the city of God. It is referred to by Isaiah as the City of Righteousness and the Faithful City." (D. Engwchit 2009)

This holy city, "as described by the prophet John, is geometric in shape....This mythology and symbolism of the utopian city—supplied by Christianity—provided the guide for the social reformers who gave birth to the modern town planning movement. Within the overcrowded Victorian slums, and within their own desires, they could see the decadence and moral corruption of Sodom, Gomorrah and Babylon. At the same time, they could see the possibility of the New Jerusalem. But for this to happen, the Babylon of London, New York, Chicago and Paris would need to be 'destroyed' and replaced with the New Jerusalem of Garden City, Broadacre City, Radiant City or City Beautiful." (D. Engwicht 2009)

In this light, it is no surprise that many of the most influential modern figures in architecture and city planning have religious roots. For instance, Frank Lloyd Wright has a grandfather who "was a Unitarian preacher. His father was a Baptist minister. Wright himself occassionally preached at the Uniterian church. His childhood was composed of 'intellectuality and genteel religion.'" (D. Engwicht 2009; R. Fishman 1982) For him, "the need for individual truth to be protected against outside attack makes the individual central in...City." (D. Engwicht 2009)

And Le Corbusier "had Calvinist mores....A religious philosopher, L'Eplattenier took the young Jeanneret under his wing and taught him how 'Rama, Krishna, Hermes, Moses, Orpheus, Pythagoras, and Jesus were all "initiates" who had penetrated the veil of matter and grasped the truth of the Spirit'.... Corbusier, recalling those days when he went with 'the Master' to sit on some mountain top, describes the experience in deeply mystical terms—'constructing a temple dedicated to nature. We shall leave the city and live in the forest at the foot of the structure....At the four corners of the temple the ceremonial fires shall be lit.'" (D. Engwicht

2009) As Hall (1988) reminded us, "Le Corbusier went as far as to declare 'The city of today is dying because it is not constructed geometrically.'" (D. Engwicht 2008)

*The Role of Rationalism*

Later, with rise of the scientific revolution, rationalism became dominant and yet remained religious in a different way, with implications for city planning.

For the ratoinalists, "emotions and irrational beliefs are not to be trusted, for they had only managed to deliver the superstitions and ignorance of the past. (The irony that it was superstition and religious beliefs that gave rise to scientific positivism escaped their attention.) The rational mind could only be trusted if it was emptied of all preconception, beliefs, and values, and disconnected from the emotions. It was believed that it was possible to extract one's self from being a player in the drama and to become an invisible observer, an observer who could move around the stage cross-examining the players about the story line without interrupting or changing the nature of the play. Or to change the metaphor, to understand how the great machine of the universe operated, it was necessary to pretend that one was no longer a cog in that machine." (D. Engwicht 2009)

Reason, then, in rationalism, the new "universal yardstick." (D. Engwicht 2009; S. Tambiah 1990). When put in the context of urban planning, "planning changed from 'a kind of craft, based on personal knowledge of a rudimentary collection of concepts about the city, into an apparently scientific activity in which vast amounts of precise information were garnered and processed in such a way that the planner could devise very sensitive systems of guidance and control.'" (D. Engwicht 2009; P. Hall 1988)

Yet, the problem with rationalism, for Engwicht (2009), is that it "is a blind leap of faith which is no different in character or nature to that undertaken in the belief system of any religion….Rationalism's great leap of faith lies in its blind acceptance that the unknown and the unknowable have no connection to that which is already known."

Or, in M. Goldberg's parlance (1985), "there is clearly a complex of variables, parameters, and interactions that we have been unable to understand or simulate, despite impressive technical modelling skills and computing machinery. Our level of ignorance about social systems is quite astounding, yet our analytical approaches to rationality assume away this ignorance outright through specification of incomplete models based on incomplete and/or often inaccurate data. Ignorance is not peripheral to our knowledge, it is central. Delimiting the bounds of ignorance is every bit as important as delimiting the bounds of our knowledge. Knowing what we do not know, as the ancient Chinese sage Lao Tzu observed, is the beginning of real knowledge: wisdom."

For Engwicht (2009), "rationalism can therefore be viewed as having the same seeds as religious fundamentalism—the need to create a simplified world, a world of order, a world of certainty. Part of the excitement of cities has always been the promise of the spontaneous, the unplanned, the random encounter and exposure to diversity. However, this also constitutes the dark side of the city and its hidden danger. For the social reformist town planning movement the danger represented by the chaotic was the danger to the moral fabric of society caused by exposure to evils such as prostitution."

The modern "rational city" can thus be traced back to the religious temptation for the "holy city" in the past.

*A Dilemma for Urban Planning*

But, if Engwicht's account of the religious roots in the history of urban planning is correct, what then is to be done in relation to urban planning? Should urban planning focus on order and purity, as conventionally done? Or should it espouse its opposite for disorder and impurity?

Alternatively, should it instead try to combine both simultaneously, as if to allure to the Chinese philosophy of "yin (passive, female, earth, negative, etc.) and yang (active, male, heaven, positive)," such that "to treat these dual banes of Western rationality [viz.., order and chaos, purity and impurity] is somehow to get in touch with them directly and become synchronous with them," as M. Goldberg (1985) and D. Engwicht (2009) would like to propose?

In *Beyond Civilization to Post-Civilization*, I already extensively analyzed why this very dualistic (or relative) worldview is equally problematic, not less than its absolute opposite (in the Christian tradition).

With this caveat in mind—my task here is to not advocate any of three possible stands, while learning from them all.

In the process, the analysis also reveals something interesting to the ontological principles in existential dialectics, and good examples are the formalness-informalness principle, the absoluteness-relativeness principle, the denseness-emptiness principle, the openness-hiddenness principle, the order-chaos principle, the symmetry-asymmetry principle, the regression-progression principle, and the same-difference principle.

For instance, in relation to the formalness-informalness principle, if there is formal logical systems (e.g., the requirement of consistency in the rationalist mode of logic), threre is also its informalist counterpart (e.g., the allowance of inconsistency and contradiction in yin-yang philosophy).

In relation to the absoluteness-relativeness principle, if there is absolute thinking (e.g., the absolute viewpoint in religious fundamentalism), there is also relative thinking too (e.g., the allowance of uncertainty in chaos theory, so that what is valid, or certain, from one standpoint may not be so from another).

In relation to the denseness-emptiness principle, if there is denseness (e.g., the problem of crowdedness in a dense environment, such that city planning is needed, even often tainted with the religious surge for order and purity), there is emptiness (e.g., the yearning for a garden of paradise without the problems associated with urban density).

In relation to the openness-hiddenness principle, if there is open revelation (e.g., the open exploration of different patterns for city planning), there is also hidden bias (e.g., the religious dogma behind the search for the "holy city").

In relation to the order-chaos principle, if there is order (e.g., the creation of order in a city planning), there is also chaos (e.g., the occurrence of chaos within a city, especially in what the Christians called the "evil city").

In relation to the symmetry-asymmetry principle, if there is symmetry (e.g., the co-existence of different religious values in urban planning), there is asymmetry (e.g., the popularity of Christian religion in medieval urban planning, and the fashionable acceptance of rationalism in modern urban planning).

In relation to the regression-progression principle, if there is progress (e.g., the design of modern buildings to confront the challenge of ever increasing urban density, as in the works of Wright and Corbusier), there is also regression (e.g., the religious bias against sexual deviance and disorder in modern urban planning).

And in relation to the same-difference principle, if there is difference in outcome (e.g., the ever new variations of city formation over time, be they about the "holy city" or the "evil city"), there is similarity in outcome too (e.g., the recurrent drive for the "great city," be it in the image of Man or that of God).

## Aesthetics and Soundscape Planning

Another way to examine urban density from the perspective of culture concerns the use of "sound landscape" (or in short, "soundscape") for aesthetics. What then is "sound landscape"?

*The Messengers of Sounds*

In an urban density, where many people reside, the design of acoustic environments becomes highly important, as it can affect their quality of life, that is, making life more comfortable or uncomfortable, for instance. In this sense, "noise pollution" in an urban density can negatively affect the quality of life.

The study of "sound landscape" is to inquire what acoustic environments are best for serving the quality of life in an urban density. Thus, "the sound at…[an urban] location certainly does not occur by chance; when people listen and notice the sounds that they create, they respond to their surroundings. The design, material and contents of the location invite activities and sounds which are associated with life and movement. The sounds say something about the landscape, the gardens and the spaces. The sounds are messengers." (P. Hedfors 2003)

These messengers can tell us something about "the distance between walls, the height of the walls, the texture of the pavements, the location of different use of land, the location of functional surfaces, the amount of vegetation, the height of the trees, water flows, etc.—these are all a part of landscape architecture and planning—they all affect the acoustic environment of any location. The sounds which accompany the design influence humans and other creatures who have the sense of hearing. These may be experienced as comfortable or uncomfortable, but they are always able to relate some amount of information." (P. Hedfors 2003)

*Sound Preferences*

In a study on "natural sounds," it was shown that some sounds seem to appeal more than some others.

For instance, L. Anderson (1983) and others showed that "'natural sources' such as insects and birds are rated positively, mechanical and engine sounds are rated negatively and human and domestic animal sounds are considered neutral or intermediate to others in a range of investigations of different outdoor environments. Sounds which might be regarded as enhancing improve wooded, natural and heavily vegetated urban settings, but not downtown or other mostly built-up sites." (P. Hedfors 2003)

In fact, another interesting finding is that there exists a correlation between the nature of sounds and the kind of urban settings. For instance, "[d]esirable sounds strongly enhance the setting, while undesirable sounds greatly detract from it. A botanical garden site and a memorial garden site are almost as sensitive to most sounds. An urban hard-surface site is relatively insensitive to the sounds, being neither greatly enhanced nor greatly marred by them." (P. Hedfors 2003)

*Technical, Aesthetic, and Other Concerns*

The danger to avoid here, of course, in designing an acoustic environment is to focus merely on "technical" issues (e.g., how to fight noise pollution) without equally considering the "aesthetic" appeal (e.g., which ways of noise pollution can better enhance the appreciation of pleasant sounds too).

Thus, "[t]he design problem is therefore one of both practical functionality and aesthetics. The aim is not just to remove the sections which are undesirable, but to weigh many factors in the formulation of solutions which are suited to the purpose. The 'right sound at the right place' enhances the character of the locations and highlights any adjustments which are made." (P. Hedfors 2003; B. Mulligan 1987; C. Harris 1998)

In fact, some scholars even propose that sound landscapes should also concern with "moral" and "spiritual" issues, to the extent that they are "biologically wholesome, socially just and spiritually rewarding….Sounds are a part of the biological, social and spiritual fibre which need to be considered if good outdoor environments are to be provided." (P. Hedfors 2003; T. Turner 1998; J. Benson 2000)

This demand then poses a formidable challenge to those "researchers" and "practitioners" working in the business of sound landscapes, since they have, for instance, not only to learn many skills and posess different fields of knowledge but also to adjudicate the conflicting demands by differernt groups.

For illustration, "[t]heir primary task is to formulate professional plans and projects; this requires a trustworthy relationship with buyers, openness to users, association with a professional ethics body, respect for skills training and efficiency of performance, among other qualities….Consequently, they are faced with almost impossible demands that they participate in skills training, keep abreast of current trends and determine when the results of investigative research are applicable. Researchers and practitioners within the field of landscape architecture and planning therefore need to develop methods and establish meeting places for the communication of results and the mutual exchange of information." (P. Hedfors 2003; I. Thompson 2000)

These conflicting demands by different groups are further complicated when researchers and practitioners in sound landscapes can be "supervised by politicians. The role came about due to the society's demand for household with natural resources and other related services." (P. Hedfors 2003)

This then means that politics can also complicate the final outcome of designing a sound landscape—especially when there is a fight among different interest groups for budget allocation and for "legal development planning" restrictions. (P. Hedfors 2003)

*The Problems with the Aesthetic Consideration of Sounds*

In the case of aesthetic consideration, as a case study of sound landscape here—the problems are not difficult to escape attention. Consider three serious problems below.

Firstly, the final decision on an aesthetic desgin for a sound landscape often do not take into account of the opinions of all the individuals affected—although on paper, everyone is supposed to have a say. And any change to an formalized aesthetic design can often be altered, however, in a reformist way, in reality.

After all, "[a]esthetic experience refers to that which affects the senses, i.e., the impressions which are generated by the senses and interpreted by the individual. The presented design proposals are increasingly tested and gradually rejected in the sketching phase….The conditions for including acoustic experiences therefore take shape gradually. The aim of the consultations in the design and planning processes should be to ascertain the ability of the concerned individuals to be comfortable in the proposed environment. The sketching phase is often passed during the consultations. For this to happen then sufficient information in the form of well-formulated supporting documents must be made available so that all concerned might understand the thought process. In practice, only serious faults are attended to after the consultations." (P. Hedfors 2003)

Secondly, the final decision on an aesthetic design for a sound landscape can also reflect the cultural bias in a society at a given time, without necessarily representing any universal agreement on the nature of sound and other senses.

For instance, "[t]he sketches and plans which are formulated during the phases appeal to visual experiences since they can be presented on paper or on a computer screen, with few exceptions. Physical models can be used in larger projects; even these contribute to the visual experience. The concerned individuals are rarely allowed to change the physical models...The gravitation towards the visual stimulates thoughts about the limitations of one culture and a mindset which is visually dominated....There is a risk that the proposals do not consider the multi-dimensional reality which is absorbed and created by merging all the senses. How else could an idea based on the impressions of one of the senses...be used to formulate an idea based on the impressions from another sense, or even complex impressions, i.e., a combination of senses?" (P. Hedfors 2003; W. Ong 1982; J. Pallasmaa 1996; D. Levin 1993)

So, how this combination of different senses is to be understood can be culture-bounded.

Thirdly, not everyone agrees with what the general sound preferences should be—although, in some cases, the disagreement may not be serious, while in other cases, it can be serious.

For instance, in one study, this time, by Herbert Kariel (1980), the gap in preference between desirable and undesirable sounds are not as wide as conventionally assumed: "Although mountaineers consider nature-related sounds more pleasant and both man- and technology-related sounds more annoying compared to other visitors, the rankings according to the degree of pleasure or annoyance are extremely close. Nature-related sounds are those of running water or waves on the shore, wind, birds, insects, and native animals. Person-related sounds are those of ordinary daily human activities such as persons talking or chopping wood and sounds made by pets. Technology-related sounds are those made by machines and equipment, such as radios, stereos, or television sets, motorboats, cars, motor or trail bikes, snowmobiles or chain saws." (P. Hedfors 2003)

The point here, for Kariel, is that for those individuals who prefer certain types of sound should be allowed to do so, just as those who prefer some other types should likewise be tolerated to choose. As Kariel (1980) thus concluded, "the fault often lies not with the actual sound level—say of a jet overflight—but rather with its intrusiveness into and incompatibility with the natural environment," which an individual may find himself into and has to resolve in his own way.

In other cases, however, this disagreement is all the more acute "for people with a diminished capacity to see and hear, but even for others in general." (P. Hedfors 2003; D. Stokols 1978).

For instance, in relation to the hearing impaired—if they "are unable to perceive the informative sounds on their own,...then they will hardly have access to the society, with the result that they feel excluded. The same feeling can hearing impaired but not deaf have if they are unable to sort the relevant sonic information out of the context. This is perhaps one of the most alarming auditory aspects of the outdoor environment with regards to social sustainability." (P. Hedfors 2003)

And in relation to others in general—Barry Truax (1984) made a distinction between "sound field" and "soundscape," in that the latter ("soundscape") is "an expression which focuses on the listener's experience of space" but the former ("sound field") "places the focus on the distribution of sound by the acoustic source. Each listener in the vicinity of an acoustic source [sound field] experiences different soundscapes." (P. Hedfors 2003)

*The Issue of Sustainability Revisited*

This critical analysis of soundscape planning in relation to aesthetic consideration reveals the major problems to achieve sustainability in soundscape.

To say that all interests and preferences of individuals in a given acoustic environment should be considered and weighted accordingly is easier said than done, especially when the reality in culture, society, nature, and the mind is vastly different from the euphoric discourse on sustainability, this time, in relation to urban density as a case study.

Yet, this analysis of soundscape planning can also teach us something about the ontological principles in existential dialectics, and good examples include the absoluteness-relativeness principle, the denseness-emptiness principle, the preciseness-vagueness principle, the simpleness-complicatedness principle, the evolution-transformation principle, the symmetry-asymmetry principle, and the regression-progression principle.

For instance, in relation to the absoluteness-relativeness principle, if there is absolute analysis (e.g., the search for universal sound preferences), there is also relative analysis (e.g., the relative experience of space as shown in different soundscapes by each listener from the same acoustic source, so that what is pleasant from one standpoint may not be so from another).

In relation to the denesness-emptiness principle, if there is denseness (e.g., the existence of urban density in the context of sound pollution), there is also emptiness (e.g., the search for "quiet space" in an urban setting).

In relation to the preciseness-vagueness principle, if there is vagueness (e.g., the vague generalization of sound preferences for all), there is also preciseness (e.g., the more precise distinction between "sound field" and "soundscape" for the finer individual experience of space).

In relation to the simpleness-complicatedness principle, if there is simple analysis (e.g., the simplistic focus on the "technical" concerns with noise pollution), there is its complicated counterpart (e.g., the more complicated analysis of how to fight noise pollution to include technical, aesthetic, social, and political problems).

In relation to the evolution-transformation principle, if there is evolution (e.g, the natural evoution of humans to hear sounds from natural objects), there is transformation (e.g., the technical invention of man-made sounds "by machines and equipment, such as radios, stereos,…television sets, motorboats, cars, motor or trail bikes, snowmobiles or chain saws"). (P. Hedfors 2003)

In relation to the symmetry-asymmetry principle, if there is symmetry (e.g., the co-existence of different sounds), there is asymmetry (e.g., the pleasantness of some sounds for certain individuals, but the unpleasantness of other sounds for them).

And in relation to the regression-progression principle, if there is progression (e.g., the implementation of soundscape planning to fight noise pollution), there is regression (e.g., the costly fight among different groups for any soundscape planning, often with the uneven appreciation of the marginal views of the disadvantaged, like the minorities, the poor, or the hearing-impaired).

## The Dysfunction of Density

In the end, one cannot help but wonder, To what extent can sustainbility (in the context of urban density as a case study here) be achieved? And to what extent can sustainability be desirable?

When the different perspectives of the mind, nature, society, and culture are factored into the larger picture, things no longer look as promising as the proponents would like us to believe. This thus constitutes the social dysfunction of urban density, in the context of sustainability.

But this is not the end of the matter yet, since the issue of sustainability can also be examined in a different (though related) way, this time, in relation to urban void—which we now turn to Chapter Three on void and its solitude to analyze.

**Table 2.1. Density and Its Crowdedness**

- **Density and the Mind**
  —Ex: Traffic Density, Road Rage, and Human Emotion
  —Ex: Urban Density and Mental Health

- **Density and Nature**
  —Ex: Ecological Complexity, and the Urban Balance with Nature

- **Density and Society**
  —Ex: Urban Pollution and Institutional Failures
  —Ex: Urban Renewal and Structural Injustice
  —Ex: Sustainable Technology and System Complexity

- **Density and Culture**
  —Ex: Urban Planning and the Religious Search for Order and Purity
  —Ex: Aesthetics and Soundscape Planning

*Notes*: The examples in the categories are solely illustrative (not exhaustive), and the comparison is relative (not absolute), nor are they necessarily mutually exclusive. And some can be easily re-classified elsewhere. As generalities, they allow exceptions.
*Source*: A summary of Ch.2 of *FPHUP*

**Table 2.2. The Demand and Supply Sides of Traffic Reduction**

- **The Demand Side**
  - —"Parking restrictions, making motor vehicle use less attractive by increasing the monetary and non-monetary costs of parking, introducing greater competition for limited city or road space. (D. Shoup 2005; H. Knoflacher 2006)
  - —"Park and ride facilities allowing parking at a distance and allowing continuation by public transport or ride sharing. Park-and-ride car parks are commonly found at metro stations, freeway entrances in suburban areas, and at the edge of smaller cities."
  - —"Reduction of road capacity to force traffic onto other travel modes. Methods include traffic calming and the shared space concept."
  - —"Road pricing, charging money for access onto a road/specific area at certain times, congestion levels or for certain road users"
  - —"Road space rationing, where regulatory restrictions prevent certain types of vehicles from driving under certain circumstances or in certain areas," with the use of "number plate restrictions" and "permits." (LEDA 2008; J. Henley 2005)

- **The Supply Side**
  - —"Adding more capacity at bottlenecks (such as by adding more lanes at the expense of hard shoulders or safety zones, or by removing local obstacles like bridge supports and widening tunnels)"
  - —"Adding more capacity over the whole of a route (generally by adding more lanes)"
  - —"Creating new routes"

*Notes*: The examples in the categories are solely illustrative (not exhaustive), and the comparison is relative (not absolute), nor are they necessarily mutually exclusive. And some can be easily re-classified elsewhere. As generalities, they allow exceptions.

*Sources*: From *Sec. 2.2.1* of *FPHUP*. For details, see WK (2008k), D. Shoup (2005), H. Knoflacher (2006), LEDA (2008), and J. Henley (2005).

**Table 2.3. Some Serious Consequences of Traffic Density**

- "Wasting time of motorists and passengers ('opportunity cost'). As a non-productive activity for most people, congestion reduces regional economic health."

- "Delays, which may result in late arrival for employment, meetings, and education, resulting in lost business, disciplinary action or other personal losses."

- "Inability to forecast travel time accurately, leading to drivers allocating more time to travel 'just in case,' and less time on productive activities."

- "Wasted fuel increases air pollution and carbon dioxide emissions (which may contribute to global warming) owing to increased idling, acceleration and braking. Increased fuel use may also in theory cause a rise in fuel costs."

- "Wear and tear on vehicles as a result of idling in traffic and frequent acceleration and braking, leading to more frequent repairs and replacements."

- "Stressed and frustrated motorists, encouraging road rage and reduced health of motorists."

*Notes*: The examples in the categories are solely illustrative (not exhaustive), and the comparison is relative (not absolute), nor are they necessarily mutually exclusive. And some can be easily re-classified elsewhere. As generalities, they allow exceptions.
*Source*: From *Sec. 2.2.1* of *FPHUP*. For details, see WK (2008k).

**Table 2.4. Opposing Views on Human Relations**

- **The Negative**
  - —*Homer*: Humans get bored of sleep and love more quickly than of war.
  - —*G.W.F.Hegel*: War can strengthen nations, but peace can corrupt them.
  - —*Friedrich Nietzsche*: The Christian virtues of humility and peacefulness are for the herds, not for the higher spirits.
  - —*St.Augustine:* War reflects the corrupt nature of human sinfulness.
  - —*Niccolo Machiavelli*: The prince needs to commit evil if necessary for power.
  - —*Thomas Hobbes*: Humans are nasty, brutish and short in the state of nature, with "a war of every man, against every man."
  - —*Konrad Lorenz*: There is some innate aggressive human nature.
  - —*Sigmund Freud*: Humans are endowed with aggressive instincts.
  - —*Heinrich von Treitschke*: To abolish war is itself immoral.
  - —*Benito Mussolini*: War is the highest expression of human power

- **The Affirmative**
  - —*Jean-Jacques Rousseau*: Humans are naturally shy and innocent but are corrupted by civilization.
  - —*Lao Tzu*: Humans are profoundly intuitive and can live a peaceful life by co-existing with nature.
  - —*Jean-Paul Sartre*: "Existence precedes essence," so there is no human nature.
  - —*Hindus*: Humans are essentially spiritual and should strive for the divine reality of the unity of all things.
  - —*Immanuel Kant*: Humans are free to obey moral laws to overcome their animality.
  - —*Jesus Christ*: If someone slaps you, turn your other cheek.
  - —*Confucius and Mencius*: Humans are essentially good and, with the right teaching of rules and duties, can make a harmonious society.

*Notes*: The examples in the categories are solely illustrative (not exhaustive), and the comparison is relative (not absolute), nor are they necessarily mutually exclusive. And some can be easily re-classified elsewhere. As generalities, they allow exceptions.
*Source*: From *Sec.3.1.3* of *FPHC*. See text for more info and references.

**Table 2.5. Four Difficulties to Obtain Ecological Knowledge**

- Firstly, a main difficulty here concerns "timing," in that a master plan is often already decided well before LEK has a chance to influence it. This then complicates the process of collecting LEK on time. In other words, "when a sketch of a new plan is introduced to the public, a certain number of basic conditions have already been determined, such as the master plan, aim of surface area to be built, and architectural vision." (J. Kohl 2005)
- Secondly, another main difficulty concerns the scope of collecting LEK from all participants involved. The difficulty here is that "[p]lanning officials inform the public concerning the availability of plan preparation material, but these information channels do not always reach all interested parties. Local residents, and other participants, may feel that the release of information is inadequate." (J. Kohl 2005)
- Thirdly, a third difficulty concerns the use of public discussion in collecting LEK. In other words, the difficulty with the use of public discussion is that "[p]ublic discussions are aimed at all interested parties, but it is rather uncommon that all participants are able to attend the forums at the same time....[S]ome participants perceive that other methods of influencing the plan are more useful than public discussions....The loudest opponents may well get their voices heard, while those who are satisfied stay silent." (J. Kohl 2005)
- And fourthly, the even more serious difficulty concerns the enormous problem of trying to understand the "complexity" of nature, especially in relation to (i) "self-organization," (ii) "connectivity," (iii) "criticality," (iv) "novelty," (v) "diversity," and (vi) "emergence," to the effect that both order and chaos are part of the nature of things. This is something that I already went to great lengths to analyze in *The Future of Complexity* (*FC*). So, the challenge here is, How can individuals collect all the info as needed, if the ecosystem in question is more complex and uncertain than what the enthusiastic urban planners would like to admit?

*Notes*: The examples in the categories are solely illustrative (not exhaustive), and the comparison is relative (not absolute), nor are they necessarily mutually exclusive. And some can be easily re-classified elsewhere. As generalities, they allow exceptions.
*Source*: From *Sec. 2.3* of *FPHUP*. For details, see J. Kohl (2005).

**Table 2.6. Three Difficulties to Use Ecological Knowledge**

- Firstly, a main difficulty here concerns the numerous (often conflicting) pieces of information on LEK to choose from. So, which ones are to be used and which ones are to be discarded? As Dennis Frenchman (2008) once observed, an recurrent challenging problem in designing and building cities is, Who are to decide what to include or exclude in the designing and building process? Also, "[r]eceived information regarding local nature may vary from opinions, feelings, and experiences to very knowledgeable species observations and well-founded perspectives on the development of urban green areas." (J. Kohl 2005)
- Secondly, another main difficulty concerns the reliability of info on LEK. For instance, "[p]lanners prefer well-founded and reliable knowledge, but they must be able to determine how reliable the knowledge source is....The interviewed planners emphasize that local knowledge must be treated critically, because it may easily contain inaccuracies....The situation becomes more difficult if the observation is several years old, because circumstances may have changed." (J. Kohl 2005)
- And thirdly, a related difficulty concerns the subjective (or emotional) nature of info on LEK from participants. In other words, the difficulty here concerns how subjective info "is valued. The interviewees often regard LEK as subjective (associated with opinions and feelings), which creates challenges in how to use the knowledge....LEK can be very emotional, because ordinary residents may not be able to argue clearly about their experiences and opinions. However, as the interviewed planners note, emotional stances have their reasons, and a good planner tries to interpret what is at the root of these emotional outbursts. Furthermore, local participants can appreciate different elements of nature in different ways....A planner, however, has to take both into consideration as a part of the planning process. The different perspective of each resident may thus become a dilemma for the planner who tries to incorporate these conflicting messages." (J. Kohl 2005)

*Notes*: The examples in the categories are solely illustrative (not exhaustive), and the comparison is relative (not absolute), nor are they necessarily mutually exclusive. And some can be easily re-classified elsewhere. As generalities, they allow exceptions.
*Source*: From *Sec. 2.3* of *FPHUP*. For details, see J. Kohl (2005) and D. Frenchman (2008).

# • PART THREE •

## *Void*

# CHAPTER 3
## *VOID AND ITS SOLITUDE*

I never found the companion that was so companionable as solitude.
—Henry David Thoreau (HC 2009)

## The Attraction of Void

The other side of density is void, which is attractive for some people, as a way to counter the problems associated with the former (urban density).

Between density and void, however, are various degrees of the combination of both. In this sense, both density and void should be treated as an approximation, in various degrees—not in an extreme (or 100%) absoluteness.

That said, the issue concerning the possibility and desirability of sustainability can be examined within the context of void in urban planning.

A comprehensive analysis of void in this sense can be provided from the four perspectives of the mind, nature, society, and culture.

In other words, this chapter is organized into four sections, in relation to (3.2) void and the mind, (3.3) void and nature, (3.4) void and society, and (3.5) void and culture—to be discussed in that order below (and summarized in *Table 3.1*).

## Void and the Mind

The nature of void in the context of urban planning from the perspective of the mind can in turn be analyzed with two good case studies here, namely, (3.2.1) indifference and urban solitude; and (3.2.2) urban homelessness and mental health—to be addressed hereafter, respectively.

## Indifference and Urban Solitude

One easy escape from urban density is to take refuge in urban void. But to what extent is this attempt successful, both in terms of possibility and desirability, from the vantage point of urban planning?

### *The Dilemma Between Solitude and Community*

To start, there is, in the absence of better terms, what I would like to call *a dilemma between solitude and community* in a city, which, however, is not hard to understand.

On the one hand, "the city" is often regarded, rightly in a way, "as a primary site for the production of community," and this is true, even when an urban population is very diverse, to the point that there is the recurrent search for "new and different bases for community" for diverse groups of urban residents—or, in short, "for subcultural formations, for affective and elective ties, for re-imagining the nature of community." (F. Tonkiss 2003)

Well, at least this is the ideal, even if not so in reality.

On the other hand, the city has never really succeeded in providing a sustainable community for all, to the extent that life in an urban density has been regarded by the critics as "isolating, anonymous, degrading of social ties, inimical to community." (F. Tonkiss 2003)

Not surprisingly, there is this ever temptation for solitude (or privacy) in an urban void, as an escape from the dreads of urban density.

### *Face-to-Face Relations vs. Side-By-Side Relations*

The temptation for solitude has to do with what Fran Tonkiss (2003) called the "conception of indifference," which is "premised less on the 'face-to-face' relations of community than on the 'side-by-side' relations of anonymity" (as summarized in *Table 3.2*).

In "face-to-face" relations, the community in question can take different social imaginations, of course, especially for "particular groups—often defined ethnically or spatially—as constituencies or objects of government, tracing a boundary around what are then produced as common interests or problems." (F. Tonkiss 2003)

In this language of community, "face-to-face" relations can provide the basis "for mobilization, for opposition, for a positioning of and claim to voice. Here, discourses of community frame an assertive rather than a defensive politics of difference, a collective agent in arguments for social and cultural recognition and the pressing of group or minority rights." (F. Tonkiss 2003; S. Benhabib 1996; A. Gutmann 1994; C. Taylor 1992).

In this sense, "face-to-face" relations represent the idea of "recognition," in the parlance of Denise Riley (2002), for those who are psychologically predisposed to be "gregarious," in that "[i]t assumes a certain sociality, a visible presence in social space." (F. Tonkiss 2003)

By contrast, in "side-by-side" relations, the individuals in question psychologically prefer "to be unseen, to be unexceptional, to be impersonal in a social field where 'differences remain unassimilated'...and strangeness a matter of fact." (F. Tonkiss 2003; I. Young 1990)

In this sense, "side-by-side" relations constitutes the desire of "loneliness and privacy," to the extent that "cities" are now regarded by these individuals as "places in which relations of non-identity are possible, tolerable, even normal. They can encourage an indifference to matters of difference that affords certain protections and allows certain freedoms. This is not simply to set indifference in the city against the various workings of community or identity, the solitude of the street versus the convivial politics of the street party; rather, it is to suggest that part of a politics of difference (important because mundane) is when differences go unremarked because unremarkable, where otherness is ordinary, where a logic of anonymity displaces one of visibility. Indifference is, in this sense, one way in which differences are lived in everyday social spaces. An 'ethics' of indifference may be (and perhaps usually is) inadvertent or unthinking, and the licence of anonymity, 'the right to be lonely' granted by the lack of concern of others, may be a temporary, uneven and unreliable freedom." (F. Tonkiss 2003)

*Mental Life and the Desire to Be Lonely*

The psychological yearning for lonely freedom in the city is the search for "the lonely liberty of knowing that no one is looking, nobody is listening. Solitude appears here as a kind of social relation, as a form of social being, rather than simply as the absence of sociality." (F. Tonkiss 2003; E. Levinas 1987; D. Riley 2002).

Georg Simmel (1997), in "The Metropolis and Mental Life," once observed that there is this "tendency to reserve among metropolitans, a reserve that shades 'not only into indifference, but, more often than we are aware...a slight aversion, a mutual strangeness and repulsion'....[These] relations of indifference or even aversion are the only feasible way of being together in a crowded city. Any one person might be too much to take, on top of everything else that is going on around you. What appears as dissociation is, in fact, the basic form of urban sociation, one that allows us to coexist with all these largely unknown others. Not interacting with others in this sense becomes a primary condition for urban social life, securing individual calm together with relative social peace." (F. Tonkiss 2003)

The desire to be lonely, for Simmel (1997a), is "psychological," in the sense that it "reinforces the fact of otherness in social life. Difference is understood here as a profoundly spatial reality, one lived over and over in the glancing encounters of the street. The mundane manoeuvres of everyday life (not making eye contact on the sidewalk, ignoring the weird intimacy of the crowded subway) play out on a microscale a wider tension between individuality and collective life. In the city, as elsewhere, individual freedom has as its correlates impersonality and anonymity," and, as Simmel thus wrote: "For here, as elsewhere, it is by no means necessary that the freedom of man be reflected in his emotional life as comfort." (Simmel, 1997; F. Tonkiss 2003)

Walter Benjamin (1986) also remarked "how, on childhood walks in the city with his mother, 'solitude appeared to me as the fit state of man.'" (F. Tonkiss 2003) And, for Louis Wirth (1995), "being with others in the city is characterized by close—sometimes unsettlingly close—physical proximity twinned with social distance. Keeping your distance in the press of upclose bodies is a special urban trick or art. People, especially women, effect it all the time in the face of accidental or unwanted physical contact in public." (F. Tonkiss 2003)

Thus lies a matter of fact, which is "typical of urban life." (F. Tonkiss 2003)

*The Death of an Urban Solitary*

To say, of course, that this matter of fact concerning urban solitude is "typical of urban life" implies that, somehow, it is desirable to have, as Fran Tonkiss (2003) was quick to advocate, albeit with a sense of ambivalence. But to what extent is urban solitude really desirable?

On the one hand, there are benefits, for sure. Louis Wirth (1995) precisely made this point, when he argued that "if urban individuals are not objects of duty or especial concern to one another, neither are they objects of antipathy or even, very much, of curiosity. Although the disaffected or disconnected nature of urban sociality might fragment strong loyalties, interests or solidarities, it should also weaken strong antagonisms, animosities or grudges." (F. Tonkiss 2003)

On the other hand, however, the search for urban solitude has its own pitfalls. The story of the death of Kathy Nguyen is a good illustration, when she died, in her urban solitude with little community nor good friends, in New York in October 2001, as "the fourth victim of anthrax in the U.S. in the period following the 11 September 2001 terrorist attacks and the first to contract anthrax by inhalation." (F. Tonkiss 2003)

Although Kathy was not a celebrity but only a solitary ordinary individual lost in urban life, her story can speak volumes of the pitfalls of urban solitude.

To start, she "arrived in the U.S. as a refugee from Vietnam in 1975....At least one version had her airlifted from the roof of the U.S. embassy in Saigon. Her known relatives were killed in Vietnam. She was married in the U.S. and then divorced. Her son died in a car accident. She had lived at the same address in the Bronx for 20 or 22 years and had worked at the Manhattan Ear, Eye and Throat Hospital for 10 years....She was 61 years old when she died." (F. Tonkiss 2003)

In her daily routines, "she would leave her one-bedroom Bronx flat in the early afternoon on most days, en-route to her late shift job as a stockroom employee at a Manhattan hospital." (F. Tonkiss 2003; P. Huffstutter 2001)

She would pass "children playing in the courtyard of her building, then on, 'Past the carwash, past the Enterprise car rental shop, past La India Mexican restaurant and up the steps at Whitlock Avenue train station, where she would board the Number 6 going south into the city.'" (F. Tonkiss 2003; P. Huffstutter 2001)

She would then go "counterflow against the commuter traffic heading from Manhattan to the outer boroughs at the end of the working day, 15 stops to 68th Street and Lexington. The newspaper accounts follow the thread of everyday movement through the city, of routine paths, glancing encounters, street anonymity." (F. Tonkiss 2003)

Behind this account of her daily routines in urban life lies "the loneliness and privacy of an individual life in the city." (F. Tonkiss 2003) In her unique case, "an absence of ethnic ties is the basis of loneliness: hers was 'a most peculiarly un-Vietnamese life—no other aunties, no kids running around, no men to smoke and drink beer and tell war stories; no men, ultimately, to protect her.' Her biography, typical if viewed as an urban life hedged about by anonymity, is seen as unusual given certain assumptions about how a properly 'ethnic' life should look: generations, shared histories, the protection of men." (F. Tonkiss 2003; J. Wenger 2001)

On top of this lies another factor which contributes to her yearning for urban solitude, since "a person such as Kathy Nguyen lives 'a subcutaneous life, where just by being Asian, an immigrant, and older, people look past you.' Indifference here is understood as aversion, disregard or neglect. It points to the way that some urban subjects are more invisible, more anonymous than others." (F. Tonkiss; E. Guillermo 2001)

So, in the end, as D. Riley (2002) aptly observed, "the admission of even occasional loneliness remains taboo" in urban density, but "to be without visible social ties [for urban solitude] is inexcusable," because "a taint of vice always clouds it." (F. Tonkiss 2003) In fact, "the special capacity of cities for privacy also allows for unconcern, neglect, disrespect." (F. Tonkiss 2003)

*In Retrospect*

This then constitutes what I earlier called "the dilemma between solitude and community." Any effort to sustain community in urban density is as problematic as the one to sustan solitude in urban void. Surely, an urban planner can choose one over the other, or the other over one, or a combination of both, but the darker sides of each remain, albeit in different degrees.

This story, tragic as it is, can also reveal something about the ontological principles in existential dialectics, and some good examples include the absoluteness-relativeness principle, the denseness-emptiness principle, the simpleness-complicatedness principle, the openness-hiddenness principle, the finiteness-transfiniteness principle, the symmetry-asymmetry principle, and the regression-progression principle.

For instance, in relation to the absoluteness-relativeness principle, if there is absolute viewpoint (e.g., the fashionable talk of building community in urban density), there is relative counterpoint (e.g., the yearning for solitude in urban void from the Others, so that what is valid from one standpoint may not be so from another).

In relation to the denseness-emptiness principle, if there is denseness (e.g., the existence of "face-to-face" relations in urban density), there is also emptiness (e.g., the existence of "side-by-side" relations in urban void).

In relation to the simpleness-complicatedness principle, if there is simplistic analysis (e.g., the death of Kathy Nguyen due to anthrax by inhalation), there is complicated account (e.g., the more complicated, detailed analysis of her death in the context of urban solitude).

In relation to the openness-hiddenness principle, there is openness (e.g., the exploration of how to build sustainble community in urban density), there is hiddenness (e.g., "the admission of even occasional loneliness remains taboo," in D. Riley's words).

In relation to the finiteness-transfiniteness principle, if there is finite number (e.g., the different individuals in urban density who can relate to the urban solitude of Kathy Nguyen), there is also transfinite number (e.g., the even greater number, maybe in millions or billions, of those residents in metropolis around the world trying to search for a community suitable to them in urban density).

In relation to the symmetry-asymmetry principle, if there is symmetry (e.g., the co-existence of both "face-to-face" relations and "side-by-side" relations in urban life), there is asymmetry (e.g., "some urban subjects are more invisible, more anonymous than others," in that "side-by-side" relations can be had for some, while the others opt for the "face-to-face" relations).

And in relation to the regression-progression principle, if there is progress (e.g., the building of different forms of community in urban density), there is regression (e.g., the emergence of urban solitude for those who do not fit in, with the subsequent consequences of "aversion, disregard or neglect").

## Urban Homelessness and Mental Health

Another way to look into urban void concerns the phenomenon of urban homelessness, with people wandering from one place to another in the street without a warm home to stay and a close community to related to—and this is all the more revealing, when examined from the vantage point of mental health.

### *Some Findings on Urban Homelessness*

A good starting point is to show some of the basic characteristics of the urban homeless.

For instance, Don Wasylenki (1998) and George Tolomiczenko found some revealing data concerning the urban homeless, this time, in the City of Toronto as a case study, as shown below (and summarized in *Table 3.3*):

- Marital Status: "72.3 percent…were single, 8.7 percent…were separated, 14.3 percent…were divorced, and 0.7 percent…were widowed. This speaks to the social isolation experienced by the individuals in the study sample."
- Education: "64.3 percent of the study sample did not complete high school."
- Income: "37.7 percent of the study sample have no income" at all.

- Encounters with the Legal System: "Between 50-60 percent of the study sample had encounters with the legal system at some point in their lives."
- Nights Spent in Police Stations or Jails in Year Prior to Interview: "People spent a lot more time in police stations or jails (30 percent ) than they did in psychiatric facilities (6 percent )."
- History of Homelessness: "The study sample showed 42 percent experiencing their first episode of homelessness, 44 percent having a pattern of chronic homelessness (three or more episodes), and 14 percent who had been homeless twice."
- Age of First Episode of Homelessness: "Homelessness is a pattern that develops relatively early in life: 131 subjects were homeless before the age of 18 years; by 30 years of age, 2/3 of the sample had experienced their first episode."
- Nights Spent on Street in Year Prior to Interview: "Over 60 percent of the sample spent time on the street, with two thirds spending more than 2 weeks on the street."
- Health Problems and Mental Illness: "People in the study sample reported chronic health problems, handicaps, disabilities, injuries and short-term illnesses. These were primarily orthopedic, respiratory and circulatory problems. While a very small proportion of the sample reported tuberculosis, it is known that tuberculosis is becoming a very significant and serious health problem in the homeless population."
- History of Childhood Abuse: "The rate of childhood sexual abuse and physical abuse reported by women and men in the homeless population was high compared with rates reported in the general population. Childhood sexual abuse: women (48.5 percent vs. 12.8 percent ); men (16 percent vs. 4 percent). Physical abuse: women (51 percent vs. 21 percent ); men (38 percent vs. 31.2 percent )."
- Substance Abuse: The use of "alcohol, marijuana and cocaine in particular) is a major problem in this population. When prevalence rates of substance abuse (drug and alcohol abuse) are added to prevalence rates of mental illness, the overall prevalence rate rises to 86 percent . Only 14 percent of the study sample had no diagnosis of either mental illness or substance abuse. The prevalence rate of alcohol and substance abuse is almost identical to the prevalence rate of mental illness; roughly two thirds of the entire homeless population. This is 4-5 times the prevalence rate in the general population. Almost everyone with a lifetime diagnosis of mental illness also had a diagnosis of substance abuse. Three quarters of the

> people in every diagnostic category of mental illness also had substance abuse disorders."

With these characteristics in mind—a tempting question to then ask here is, To what extent is urban homelessness related to mental illness, although there are other causes too?

*Mental Illness and Urban Homelessness*

In the literature on homelessness and mental health, a common correlation found in some research projects concerns the relationship between urban homelessness and mental illness, although there are other causes too.

For instance, David Michaels (1992) and others in a project titled. "Homelessness and Indicators of Mental Illness Among Inmates in New York City's Correctional System" used data drawn "from three cross-sectional samples of inmates in the New York City correctional system (N=299, 236, and 151)...to determine the prevalence of homelessness among detainees." (NCH 2008)

Some interesting findings in the project showed that "homelessness was strongly associated with mental illness: 50 percent of those who had ever been homeless during the past three years responded positively to at least one mental illness screening question, compared with 25 percent of the never-homeless inmates. More than a third of the ever-homeless group had received mental health treatment, compared with a fifth of those who had never been homeless." (D. Michaels 1992)

In another report, this time, a 2007 survey for the U.S. Conference of Mayors, it was shown that, "of the 23 cities surveyed, 7.9% of the homeless population of individuals in a family suffer from some type of mental illness. Additionally, 22.4% of the homeless individuals in this survey have a mental illness." (NCH 2008)

A different study by Don Wasylenki (1998) and George Tolomiczenko also found "the prevalence of lifetime diagnoses of mental illness in the homeless population. Two thirds (approximately 66 percent ) of the homeless population have a lifetime diagnosis of mental illness, which is 2-3 times the prevalence rate of the general population. Depression is the mental illness most reported, understandable given the extent of unfortunate circumstances in these people's lives."

In fact, "a sub-sample (29 percent ) met the criteria for anti-social personality disorder, often in addition to an Axis I diagnosis (diagnosis of psychotic disorder, depression, or post-traumatic stress disorder). About 25 percent of the sample had received psychiatric outpatient services in the year leading up to the interview, and less than 20 percent had received any kind of services for substance abuse problems." (D. Wasylenki 1998)

To put things in a proper perspective, this is so, even though "an average of 16% of the single adult homeless population suffers from some form of severe and persistent mental illness," and "22% of the American population [that is, about 44 millions, roughly speaking] suffers from a mental illness." (NCH 2008; NRTC 2003)

*The Other Causes of Urban Homelessness*

Then, a critical question to follow up is, Besides the one of mental illness, are there other causes of urban homelessness?

Paula Goering (1998) and Tammy Morrell-Bellai suggested other reasons, on top of mental illness, as shown below (and summarized in *Table 3.4*):

- "Loss of job or having inadequate income to pay rent"
- "Interpersonal conflict and divorce"
- "Eviction"
- "Drug and/or alcohol abuse"
- "Mental illness"

Now, the disagreement among researchers here concerns the weight of explanatory power to be given to each of the factors. Some say that mental illness is a minor factor, whereas others give it some more weight.

Goering (1998) and Morrell-Bellai also showed some common "stresses they [the homeless] had experienced in the year prior to their first becoming homeless," as shown below (and summarized in *Table 3.4*):

- "Death of a loved one"
- "Legal problems"
- "Illness of a loved one"
- "Break up of a relationship"
- "Debt"
- "Physical abuse"
- "Robbed"
- "Reduction of benefits"

In addition to the critical question concerning the weight of explanatroy power to be assigned to each of these stresses—it is even more important to ask why some individuals end up experiencing stresses like this in urban life and subsequently becoming homeless, while others simply bounce back from experiencing life stresses like this in urban life and later become wiser and more successful in life.

Wasylenki (1998), for instance, argued that there are two sides of the debate to be considered, in that, while "homelessness is indeed…a social problem,…there are also individual factors"; after all, "in the mental health literature about homelessness, they tend to repeat that 30-50 percent of people who are homeless have a mental illness, and that mental illness must be a major causative factor of homelessness."

And, in a more comprehensive way, in *Beyond Nature and Nurture* (*BNN*), I already went to great lengths to explain how and why some groups are more successful than others in coping with life challenges—from the four perspetives of the mind, nature, society, and culture.

*The Impact of Mental Illness on the Homeless*

Although the causes of urban homelessness can be multiple, let's consider more deeply the impact of mental illness on the homeless, as a case study in this section.

A report by the National Coalition for the Homeless described in detail how "[m]ental disorders prevent people from carrying out essential aspects of daily life, such as selfcare, household management and interpersonal relationships. Homeless people with mental disorders remain homeless for longer periods of time and have less contact with family and friends." (NCH 2008)

Even worse, "[m]any of the mentally ill (especially those with severe disorders such as schizophrenia, bipolar disorder, and major depression), both the homeless and others, often misinterpret the guidance of others and react irrationally because of their condition(s). This pushes away friends and family and other caregivers occasionally leading to homelessness or a longer a period of homelessness." (NCH 2008; NHCHC 2000)

Consequently, they become dysfunctional in relation to the larger society: "The mentally ill homeless population encounters more barriers to employment, tend to be in poorer physical health, and have more contact with the legal system than homeless people who do not suffer from mental disorder. All people with mental disorders, including those who are homeless, require ongoing access to a full range of treatment and rehabilitation services to lessen the impairment and disruption produced by their condition." (NCH 2008)

*The Danger of Two Extreme Views*

The analysis of the debate here is to guard us against two extreme views often found in the literature on urban homelessness.

One extreme view is to blame everything, or most of it, on culture and society, without due consideration of individual bio-psychological factors. And the other extreme view is to blame everything, or most of it, on individual bio-psychological makeup (like mental illness), without taking into account of social and cultural factors.

As indicated before, I already showed, in *Beyond Nature and Nurture* (*BNN*), that both nature and nurture closely interact with each other to produce the achievement gap between those who make it in society and those who don't and later become homeless. In other words, neither of the two extreme views is correct.

In addition, this analysis of mental health and urban homelessness within the context of urban void reveals something relevant to the ontological principles in existential dialectics, and some good examples are the absoluteness-relativeness principle, the predictability-unpredictability principle, the explicability-unexplicability principle, the denseness-emptiness principle, the simpleness-complicatedness principle, the openness-hiddenness principle, the slowness-quickness principle, the symmetry-asymmetry principle, and the regression-progression principle.

For instance, in relation to the absoluteness-relativeness principle, if there is absolute viewpoint (e.g., the blaming by some social workers on culture and society for for the problem of urban homelessness, as revealed in the Workshop of the Mental Health Policy Research Group, where some argued that urban homelessness is a "social," not "individual," problem), there is also relative standpoint (e.g., the response by Wasylenki, to show that the extreme view of blaming society and culture is too absolute, as both social and individual factors are important for the study of urban homelessness, such that things are more relative, or what is valid from one standpoint may not be so from another).

In relation to the predictability-unpredictability principle, if there is predictability (e.g., the higher chance to become homeless for individuals who suffer from mental illness), there is also unpredictability (e.g., the inability to predict urban homeless on the basis of mental illness alone, since not every homeless individual is mentally sick).

In relation to the explicability-inexplicability principle, if there is explicabililty (e.g., mental illness as a cause of urban homelessness), there is also inexplicability (e.g., the inexplicability of why some individuals who are not mentally ill but become homeless, if mental illness is solely considered in the explanation).

In relation to the densenss-emptiness principle, if there is urban density (e.g., the formation of close community for urban residents), there is likewise urban void (e.g., the wandering of the homeless in the street without home nor community).

In relation to the simplenss-complicatedness principle, if there is simplistic analysis (e.g., the simplistic focus on substance abuse and mental illness in addressing the characteristics of urban homelessness), there is complicated analysis too (e.g., the more complicated analysis of other main characteristics of urban homelessness, such as education, income, marital status, history of homelessness, childhood abuse, and the like).

In relation to the openness-hiddenness principle, if there is openness (e.g., the exploration of the causes of urban homelessness), there is hiddenness (e.g., the reluctance by many researchers, due to the fear of political incorrectness, to emphasize mental illness in the analysis of urban homelessness).

In relation to the slowness-quickness principle, if there is slowness (e.g., the slow recovery of some individuals after experiencing life stresses in urban life and later becoming homeless), there is also quickness (e.g., the ability of other individuals who are able to bounce back from experiencing life stresses in urban life and later become wiser and more successful in life).

In relation to the symmetry-asymmetry principle, if there is symmetry (e.g., the co-existence of different views on the causes of urban homelessness), there is asymmetry (e.g., the political correctness in our time to not emphasize too much on the bio-psychological causes of urban homelessness, as opposed to the older days when the homeless could not more easily regarded as suffering from madness).

And in relation to the regression-progression principle, if there is progress (e.g., the numerous efforts by both federal and local governments to provide different aids to the homeless, like shelters and psychiatric services), there is regression (e.g., the persistence of urban homelessness over time, due to bio-psychological causes like mental illness and others).

## Void and Nature

Another way to explore urban void in relation to sustainability in urban planning can be had from the perspective of nature.

Two good cases in point concerns (3.3.1) deforestation and urban destructiveness, and (3.3.2) urban open space and its controversy—to be analyzed in what follows, respectively.

## Deforestation and Urban Destructiveness

The examination of deforestation can say much of urban destructinevness in creating void, this time, in relation to nature.

*The Two Sides of Deforestation*

The word "deforestation" refers to "the conversion of (forest)ed areas to non-forested land, for uses such as: pasture, urban use, logging purposes, and can result in arid land and wastelands." (WK 2009) In this sense, deforestation is essentially "anthropogenic" (or human-caused), althoguh it can be "naturally" caused too.

In relation to "naturally" caused deforestation, "forests disappear naturally as a result of broad climate change, fire, hurricanes or other disturbances." (WK 2009)

But in relation to "anthropogenic" deforestation, "most deforestation in the past 40,000 years has been anthropogenic. Human induced deforestation may be accidental,...as in the case of forests in Europe adversely affected by acid rain. Improperly applied logging, fuelwood collection, fire management or grazing can also lead to unintentional deforestation. However, most anthropogenic deforestation is deliberate. The scale of deforestation can be observed with the satellite viewing program Google Earth." (WK 2009; B. Tomas 1997; J. Allen 1985)

*Different Causes of Anhtropogenic Deforestation*

In the case of anthropogenic deforestation, there are multiple causes, and good examples are shown below (and summarized in *Table 3.5*): (WK 2009; H. Geist 2002; SD 2008)

- "Farming" (e.g., "slash-and-burn")
- "Cattle ranching" (a form of animal husbandry)
- "Wood extraction" ("logging")
- "Infrastructure expansion"
- "Mining"
- "Population growth"
- "Urbanization"

Yet, there are three qualifications to be made here, lest any misunderstanding occurs.

Firstly, the list of the causes of anthropogenic deforestation can be longer, so the examples as cited above are illustrative only.

Secondly, the causes can also be mutually related. For instance, "logging (wood extraction) or mining requires roads to transport the timber (infrastructure expansion) and farmers use these roads to move into previously unreachable areas of forest (agricultural expansion). The ultimate cause of most deforestation is increased food production. Cattle, permanent crops, shifting cultivation and colonization are all equally important to global tropical deforestation." (WK 2009; H. Geist 2002)

And thirdly, there is no robust general "agreement on what causes deforestation." (WK 2009)

For instance, "logging may be a direct source of deforestation in some areas and have no effect or be at worst an indirect source in others due to logging roads enabling easier access for farmers wanting to clear the forest; experts do not agree on whether logging is an important contributor to global deforestation and some believe that logging makes considerable contribution to reducing deforestation because in developing countries logging reserves are far larger than nature reserves." (WK 2009; A. Angelsen 1999; W. Laurance 1999)

And some scholars also questioned the "claim that population growth drives deforestation," which "is weak and based on flawed data," since "population increase due to high fertility rates…[is] a primary driver of tropical deforestation in only 8% of cases. The FAO states that the global deforestation rate is unrelated to human population growth rate, rather it is the result of lack of technological advancement and inefficient governance." (WK 2009; A. Marcoux 2000; A. Angelsen 1999; H. Geist 2002)

*The Unsustainability of Anthropogenic Deforestation*

With these caveats in mind—suppose, for the sake of argument, the need to increase food production is the ultimate cause of anthropogeneic deforestation. Now a more important issue concerns whether or not anthropogenic deforestation is sustainable.

The evidence, so far, however, is negative, even though anthropogenic deforestation has helped increasing food production worldwide. But at what costs? And how long can this continue?

For instance, on the one hand, "in open forest and woodland communities food production can be increased by orders of magnitude when trees are removed." (WK 2009)

But, on the other hand, "rain forests can not support human populations at all because the food resources are too scattered," and "[t]he planet could not support current population and current living standards." (WK 2009)

This is all the more serious, when there is the global divide between rich countries and poor countries (which suffer the most).

In other words, "[f]rom the perspective of the developing world, the benefits of forest as carbon sinks or biodiversity reserves go primarily to richer developed nations and there is insufficient compensation for these services....Developing countries feel that some countries in the developed world, such as the United States of America, cut down their forests centuries ago and benefited greatly from this deforestation and that it is hypocritical to deny developing countries the same opportunities: that the poor shouldn't have to bear the cost of preservation when the rich created the problem." (WK 2009)

The challenge here to urban planners is therefore two-fold, namely, how to resolve the conflicting views between the rich and the poor, and how to deal with the curent huge world population (especially in the less developed world) and the current high living standards (especially in the developed world), both of which the planet could not sustain in the long-run.

It is easier said than done, of course, and thus the possibility of sustainability and its desirablity are essentially contestable.

*The Impacts of Anthropogenic Deforestation*

To see why anthropogeneic deforestation is not sustainable, just consider the following impacts on nature (and for that matter, society), be they (a) atmospheric, (b) hyrdologic, (c) soil, (d) ecological, and (e) economic—as summarized in *Table 3.6*:

(a) Atmospheric

Deforestation can have a serious impact on the atmosphere. (WK 2009; NASA 2004; R. Mweninguwe 2005; SD 2009 & 2009a; C. Matthews 2009; P. Fearnside1 2004; FC 2008; K. Denman 2007; I. Prentice 2009)

After all, deforestation, be it human-caused or naturally caused, "is a contributor to global warming, and is often cited as one of the major causes of the enhanced greenhouse effect." (WK 2009)

As an illustration, "tropical deforestation is responsible for approximately 20% of world greenhouse gas emissions. According to the Intergovernmental Panel on Climate Change, deforestation, mainly in tropical areas, account for up to one-third of total anthropogenic carbon dioxide emissions." (WK 2009)

In the state of nature without deforestation, "trees and other plants," normally, "remove carbon (in the form of carbon dioxide) from the atmosphere during the process of photosynthesis and release it back into the atmosphere during normal respiration. Only when actively growing can a tree or forest remove carbon over an annual or longer timeframe." (WK 2009)

But when deforestation occurs, "the decay and burning of wood," for instance, "releases much of this stored carbon back to the atmosphere. In order for forests to take up carbon, the wood must be harvested and turned into long-lived products and trees must be re-planted. Deforestation may cause carbon stores held in soil to be released." (WK 2009)

(b) Hydrologic

Deforestation has an impact not only on the atmosphere but also on the water cycle.

For instance, "trees extract groundwater through their roots and release it into the atmosphere. When part of a forest is removed, the trees no longer evaporate away this water, resulting in a much drier climate." (WK 2009)

Also, "shrinking forest cover lessens the landscape's capacity to intercept, retain and transpire precipitation. Instead of trapping precipitation, which then percolates to groundwater systems, deforested areas become sources of surface water runoff, which moves much faster than subsurface flows. That quicker transport of surface water can translate into flash flooding and more localized floods than would occur with the forest cover." (WK 2009)

In the end, "the presence or absence of trees can change the quantity of water on the surface, in the soil or groundwater, or in the atmosphere. This in turn changes erosion rates and the availability of water for either ecosystem functions or human services." (WK 2009)

(c) Soil

Deforestation can also affect the quality of soil.

In the state of nature, "undisturbed forest has very low rates of soil loss, approximately 2 metric tons per square kilometre (6 short tons per square mile)," but "deforestation generally increases rates of soil erosion, by increasing the amount of runoff and reducing the protection of the soil from tree litter….Forestry operations themselves also increase erosion through the development of roads and the use of mechanized equipment." (WK 2009)

A classic example concerns the Yellow River of China, in that "China's Loess Plateau was cleared of forest millennia ago. Since then it has been eroding, creating dramatic incised valleys, and providing the sediment that gives the Yellow River its yellow color and that causes the flooding of the river in the lower reaches (hence the river's nickname 'China's sorrow')." (WK 2009)

(d) Ecological

Deforestation can also affect the eco-system, in decreasing biodiversity. (WK 2009; S. Nilsson 2009)

As an illustration, "the removal or destruction of areas of forest cover has resulted in a degraded environment with reduced biodiversity. Forests support biodiversity, providing habitat for wildlife; moreover, forests foster medicinal conservation. With forest biotopes being irreplaceable source of new drugs (such as taxol), deforestation can destroy genetic variations (such as crop resistance) irretrievably." (WK 2009; J. Stock 2009; SD 2009b; B. Fruth 2009; BM 2007)

This is all the more serious in certain areas of the earth, like the tropical rainforests, which "are the most diverse ecosystems on earth," and "about 80% of the world's known biodiversity could be found in tropical rainforests," just as any "removal or destruction of significant areas of forest cover has resulted in a degraded environment with reduced biodiversity." (WK 2009; BBC 2009 & 2009a; TQ 2009; M. Mogato 2008; R. Butler 2007)

(e) Economic

Deforestation also has economic impacts.

For instance, a report at the Convention on Biological Diversity (CBD) meeting in Bonn, Germany showed that the "[d]amage to forests and other aspects of nature could halve living standards for the world's poor and reduce global GDP by about 7% by 2050….Today, developed countries continue to utilize timber for building houses, and wood pulp for paper. In developing countries almost three billion people rely on wood for heating and cooking." (WK 2009; AAAS 2009; R. Black 2008)

More often than not, "[s]hort-term economic gains made by conversion of forest to agriculture, or over-exploitation of wood products, typically leads to loss of long-term income and long term biological productivity (hence reduction in nature's services). West Africa, Madagascar, Southeast Asia and many other regions have experienced lower revenue because of declining timber harvests. Illegal logging causes billions of dollars of losses to national economies annually." (WK 2009; R. Butler 2007)

*A Reflection*

Yet, not everyone agrees on the impacts of deforestation as described above.

As an illustration, the impact of deforestation on biodiversity has been questioned, since "scientific understanding of the process of extinction is insufficient to accurately to make predictions about the impact of deforestation on biodiversity. Most predictions of forestry related biodiversity loss are based on species-area models, with an underlying assumption that as forest declines species diversity will decline similarly. However, many such models have been proven to be wrong and loss of habitat does not necessarily lead to large scale loss of species. Species-area models are known to overpredict the number of species known to be threatened in areas where actual deforestation is ongoing, and greatly overpredict the number of threatened species that are widespread." (WK 2009; S. Pimm1995)

With this caveat in mind, however, this reflection, together with the analysis of deforestation as provided above, can also teach us something important about the ontological principles in existential dialectics, and good examples include the partiality-totality principle, the predictability-unpredictability principle, the explicability-inexplicabilty principle, the denseness-emptiness principle, the simpleness-complicatedness principle, the expansion-contraction principle, the evolution-transformation principle, the symmetry-asymmetry principle, and the regression-progression principle.

For instance, in relation to the partiality-totality principle, if there is partial analysis (e.g., the engagement in logging from the individual self-interest of a wood-extraction company), there is also totalistic concern (e.g., the tragedy of the commons, as in deforestation, when looked from the larger perspective of nature and society).

In relation to the predictability-unpredictability principle, if there is predictability (e.g., the prediction of soil erosion, with the increase of deforestation), there is unpredictability (e.g., the inability to make accurate "predictions about the impact of deforestation on biodiversity").

In relation to the explicability-inexplicability principle, if there is explicability (e.g., cattle ranching as a cause of anthropogenic deforestation), there is also inexplicability (e.g., the lack of general agreement on the specific nature of the causes of deforestation, with different scholars offering different viewpoints).

In relation to the denseness-emptiness principle, if there is denseness (e.g., the increase of urban density, and its subsequent demand for more food production), there is also emptiness (e.g., the occurrence of deforestation and its subsequent desertification and other consequences, which make the affected areas unhabitable).

In relation to the simpleness-complicatedness principle, if there is simplistic analysis (e.g., the account for the individual causes of deforestation), there is complicated counterpart (e.g., the need to look into the more complicated interrelations among the different causes of deforestation).

In relation to the expansion-contraction principle, if there is expansion (e.g, the expansion of urban areas into more and more previously forested ones), there is also contraction (e.g., the increase of deforestation which renders some desertified areas unsuitable for human habitation).

In relation to the evolution-transformation principle, if there is evolution (e.g., the natural causes of deforestation), there is transformation (e.g., the anthropogenic causes of deforestation, especially with the use of technology for mining, infrastructure construction, and farming, for instance).

In relation to the symmetry-asymmetry principle, if there is symmetry (e.g., the co-existence of natural and anthropogenic deforestation), there is also asymmetry (e.g., the more serious impact of natural deforestation in antiquity, but the more serious scale of anthropogenic deforestation in our time).

And in relation to the regression-progression principle, if there is progress (e.g., the increase of food production worldwide by way of deforestation), there is also regression (e.g, the costly impacts of deforestation, be they atmospheric, hydrologic, soil, economic, and ecological).

## Urban Open Space and Its Controversy

A different analysis of urban void can be provided, this time, in relation to the phenomenon of "urban open space" as a case study here, in the conext of the possiblity and desirability of sustainiability.

### *A Definition*

The term "urban open space," in one conventional definition, refers to "land-use planning to define areas of 'parks,' 'green spaces,' and other open areas. The landscape of urban open spaces can range from playing fields to highly maintained environments to relatively natural landscapes." (WK 2009a)

In general, these areas (for urban void) are preserved, so that "development [for urban density] is indefinitly set aside." (WK 2009b)

And, "in almost all instances, the space referred to by the term is, in fact, green space. However, there are examples of urban green space which, though not publicly owned/regulated, are still considered urban open space." (WK 2009a)

*Main Functions of Urban Open Space*

But what are the functions of urban open space? Urban open space exists "to provide 'aesthetic and psychological relief from urban development,'" together with other benefits.

In one classification, its functions can be dividied into four main categories, which are, namely, (a) recreational, (b) ecological, (c) aesthetic, and (d) medical—to be analyzed hereafter (and summarized in *Table 3.7*). (WK 2009a & L. Springgate 2008)

Of course, there can be other functions too, as some scholars propose alternative but more complicated classifications of the functions of urban open spaces, which can be "economic," "cultural," "civic," "contemplative," etc. (WK 2009a; BMR 1971; M. Eysenbach 2008; D. Berry 1976)

With this qualification in mind—the four functions as described hereafter are the most often cited and serve the illustrative purpose at hand.

(a) Recreational

Urban open space is important, in that it can provide recreation as a way to reduce life stress in urban density.

There are two main types of recreation in urban open space, in that it can be "active" (e.g., "organized sports and individual exercise") or "passive" (e.g., "being in the open space"). (WK 2009a)

In this way, "[t]ime spent in an urban open space for recreation offers a reprieve from the urban environment." (WK 2009a)

(b) Ecological

The second function of urban open space is "the conservation of nature in an urban environment." (WK 2009a)

This focus on ecology in urban open space is to enhance our awareness "of the balance of nature, of natural processes and of man's place in and affect on nature—i.e., 'ecological awareness'....As human live more and more in man-made surroundings—i.e., cities—he risks harming himself by building and acting in ignorance of natural processes." (WK 2009a)

In addition, "[b]eyond this man-nature benefit, urban open spaces also serve as islands of nature, promoting biodiversity and providing a home for natural species in environments that are otherwise uninhabitable due to city development." (WK 2009a)

(c) Aesthetic

The third function of urban oepn space is to create "aesthetic value," which means that "[p]eople enjoy viewing nature, especially when it is otherwise extensively deprived, as is the case in urban environments.

Therefore, open space offers the value of 'substituting gray infrastructure.'" (WK 2009a; M. Eysenbach 2008)

(d) Medical

And the fourth function of urban open space is to "offer health benefits to city residents through exposure to a natural environment." (WK 2009a)

For instance, Richard Louv (2006) in *Last Child in the Woods* told us "the positive effects nature in general has on children, even when experienced in small islands of green within a city. Urban open spaces offer citizens relief from the strains of urban environments and everyday demands. That respite can come in the form of a walk or run, time spent sitting or reading, watching the birds, essentially any time spent in the natural environment the open space offers." (WK 2009a)

In fact, "research shows that when open spaces are attractive and accessible, people are more likely to engage in physical activity, which has obvious inherent health benefits." (WK 2009a; T. Hartig 2007)

*The Issue of Sustainability*

But is urban open space sustainable, in the long run? There are two specific issues here, namely, in relation to (a) the possibility of sustainability and (b) the desirability of sustainability, as summarized in *Table 3.8*).

(a) The Possibility of Sustainability

On the positive side, urban open space can be sustained in some ways.

As an illustration, in the context of biodiversity, "some green spaces maintain a natural environment with a native and self-sustaining ecosystem. Depending on factors such as the location of the city and the location of the space within the city, this natural open space may be a grassy field, woodland, or something aquatic such as a stream, swamp, pond or lake. Other areas may be more heavily influenced by its purpose and use. Examples of open space that would match this description are playing fields, gardens, or imposed ecosystems." (WK 2009a)

On the negative side, however, there are chronic problems which make life difficult for sustainability. Consider three examples below.

Firstly, not all species can survive in an urban open space. For instance, "species which are most often able to co-exist with man in an urban setting are usually those which 'are able to reproduce rapidly and to take advantage of transitory conditions or to evolve varieties suited to the urban situation.' Therefore, larger urban open spaces, especially those with various types of environments, are more likely to support a diverse ecosystem." (WK 2009a; D. Gill 1973)

But then, this requires a lot of resources (which are often not available) to fund the construction of "large urban open spaces which rely on a natural local ecosystem…in terms of maintaining a balanced biodiversity." (WK 2009a)

In addition, larger urban open spaces often pose another challenge, in that, to be sustainble, they must be properly "established and managed primarily to benefit natural wildlife populations in order that they may function as regional reservoirs." (WK 2009a; D. Gill 1972) It is easier said than done, however, epsecially when conflicting interests among different social groups render general agreement difficult to achieve.

Secondly, the increase of urban population over time often competes with urban open spaces (which are not unlimited) for different uses, especially when the problems involve "overcrowding, poverty, squalor, ill-health, lack of morals and morale, and so on." (WK 2009a)

And thirdly, urban open spaces can evolve over time into different segments, such that it is difficult to integrate them all in a sustainable urban environment.

This problem of "segmentation" is not trivial, as it "was particularly prominent in America during the twentieth century. Since the 'end-of-the-turn-of-the-century "romantic" stream valley park systems, open space designers have been concerned with guiding, containing or separating urban growth, distributing recreation, and/or producing scenic amenity, mostly within the framework of geometric abstractions.'" (WK 2009a; W. Roberts 1970)

By the 1990's, "urban open spaces took a path similar to that of parks, following the modernization trend of segmentation and specialization of areas. A[nd] modernity stressed 'increased efficiency, quantifiablity, predictability, and control… In concert with the additional social divisions'… open spaces grew more specific in purpose. Perhaps this increase in division of social classes' use of open space, demonstrated by the segmentation of the spaces, displays a situation similar to the privatization of London parks in the eighteenth century which displayed a desire to make classes more distinct." (WK 2009a; T. Young 1995)

Therefore, behind the problem of segmentation in urban open spaces lies the conflict of interests among different social groups.

(b) The Desirability of Sustainability

So, this brings us to the issue concerning the desirability of sustainability.

On the one hand, the sustainability of urban open spaces is desirable, for the simple reason (as discussed in the previous sub-section), in that urban open spaces serve some essential functions such as recreation, ecology, aesthetics, and public health.

On the other hand, however, there are chronic problems which make the sustainability of urban open spaces undesirable. Consider four examples below.

Firstly, urban open spaces, however well-intentioned as the urban planners are in designing them, often end up preserving "the middle class status quo" and benefiting its hidden interests. (WK 2009a)

As an illustration, consider the construction of "green belt," which "is basically an invisible line that goes around a certain area, stopping people building on there so that some of the wild and agricultural land can be saved." (WK 2009c)

Tim Harford (2006), for example, in his book on why the rich are rich and the poor are poor, argued that "a green belt is created by residents to preserve the bourgeois status quo of those already living within the zone, and especially the advantage of landlords who profit from a scarcity of housing." (WK 2009c)

In his view, although "the stated motivation and benefits of the green belt are well-intentioned (public health, environment)," the benefits are not what they are often claimed to be, since "only a small fraction of the population ever sets foot on the green belt for leisure purposes,...and...a green belt is not strongly causally linked to clean air and water....Rather, the ultimate result of the decision to green belt a city is to maintain the middle class status quo, thus exacerbating high housing prices by concentrating demand within the zone and stifling competitive forces in general." (WK 2009c)

In another study, this time, with the analysis of "public spaces" as a case study, researchers showed that urban open spaces often have "restrictions," such that "public spaces are managed to exclude certain groups—specifically homeless people and young people. Measures are taken to make the public space less attractive to them, including the removal or design of benches to restrict their use for sleeping and resting, restricting access to certain times, locking indoor/enclosed areas. Police forces are sometimes involved in moving 'unwanted' members of the public from public spaces. In fact, by not being provided suitable access, disabled people are implicitly excluded from some spaces." (WK 2009d; NCH 2004; B. D'Agostino 2005)

In addition, the "privatization" of some urban open spaces over the years have greatly benefited the corporate sector, at the expense of the less-to-do. For instance, "beginning roughly in the 1960s, the wholesale privatization of public space (especially in urban centers)...has faced criticism from citizen groups such as the Open Spaces Society. Private-public partnerships have taken significant control of public parks and playgrounds through conservancy groups set up to manage what is considered unmanageable by public agencies. Corporate sponsorship of public leisure areas is ubiquitous, giving open space to the public in exchange for higher air rights. This facilitates the construction of taller buildings with private parks; accessible only to those deemed fit." (WK 2009d)

Secondly, urban open spaces can often constitute serious barriers to urban expansion, which then lead to urban sprawl, with subsequent problems which are undesirable and, for that matter, unsustainable too.

In other words, in the current example of greenbelt, "it might spur the growth of areas much further away from the city core than if it had not existed, thereby actually increasing urban sprawl." (WK 2009c)

Good examples "commonly cited are the Ottawa suburbs of Kanata and Orleans, both of which are outside the city's greenbelt, and are currently undergoing explosive growth....This can lead to other problems, as residents of these areas have further to commute to work (if they seek employment in city) and little access to public transport. It also means people will commute through the green belt, an area not designed to cope with high levels of transportation. Not only is the merit of a green belt apparently subverted, but the green belt may heighten the problem and make the city unsustainable." (WK 2009c)

The critics are therefore quick to point out that "the greenbelts actually defeat their own stated objective of saving the countryside and open spaces. By preventing existing towns and cities from extending normally and organically, they result in more land-extensive housing developments further out—i.e., the establishment beyond the greenbelts of new communities with lower building densities, their own built infrastructure and other facilities, and greater dependence on cars and commuting, etc. Meanwhile, valuable urban green space and brownfield sites best suited to industry and commerce are lost in existing conurbations as more and more new housing is crammed into them." (WK 2009c)

And thirdly, urban open spaces can evolve over time into something else altogether, which was not what the original urban planning was for.

For instance, "there are many examples whereby the actual effect of green belts is to act as a land reserve for future freeways and other highways. Examples include sections of the 407 highway north of Toronto and the Hunt Club Rd / Richmond Rd. south of Ottawa. Whether they are originally planned as such, or the result of a newer administration taking advantage of land that was left available by its predecessors is debatable." (WK 2009c)

*A Challenge for Thought*

This analysis of sustainability, in the context of urban open space, in relation to both its possibility and desirability, has the virtue to provide a reality check, so as to caution those who are often mislead, even if unintentionnaly, by the euphoric discourse on sustainable urban planning in our time.

Sustainability is not necessarily possible and desirable, to the extent that enthusiastic urban planners would like us to believe.

In the end, the analysis also reveals something important to the ontological principles in existential dialectics, and good examples are the absoluteness-relativeness principle, the partiality-totality principle, the predictability-unpredictability principle, the denseness-emptiness principle, the simpleness-complicatedness principle, the openness-hiddenness principle, the expansion-contraction principle, the evolution-transformation principle, the symmetry-asymmetry principle, and the regression-progression principle.

For instance, in relation to the absoluteness-relativeness principle, if there is absolute thinking (e.g., the euphoric discourse on urban open spaces), there is relative counterpart (e.g., the relative analysis of class bias in urban open spaces, such that what is valid for the middle class may not be so for the homeless).

In relation to the partiality-totality principle, if there is partial viewpoint (e.g., the local design of each urban open space), there is holistic analysis (e.g., the emergence of "segmentation" over time from the overall picture, as more urban open spaces are built).

In relation to the predictability-unpredictability principle, if there is predictability (e.g., the predictable benefits of building urban open spaces), there is also unpredictability (e.g., different unintented problems over time to haunt any well-intentioned design of urban open spaces).

In relation to the denseness-emptiness principle, if there is denseness (e.g., urban expansion with the increase of urban population), there is emptiness (e.g., the need for urban open spaces as an escape from life stress in urban density).

In relation to the simpleness-complicatedness principle, if there is simple analysis (e.g., the simple illustration of four functions of urban open spaces), there is also complicated counterpart (e.g., alternative but more complicated classifications of the functions of urban open spaces, to include, say, "economic," "cultural," "civic," "contemplative," etc.).

In relation to the openness-hiddenness principle, if there is openness (e.g., the exploration of the functions of urban open spaces for all residents in a city), there is hiddenness (e.g., the hidden class bias in the construction of urban open spaces).

In relation to the expansion-contraction principle, if there is expansion (e.g., the construction of larger urban open spaces), there is contraction (e.g., the barriers to urban expansion with the construction of more urban open spaces).

In relation to the evolution-transformation principle, if there is evolution (e.g., the natural evolution of humans to live in caves with nature, in much earlier times of history), there is transformation (e.g, the use of technological means to construct urban open spaces in our times).

In relation to the symmetry-asymmetry principle, if there is symmetry (e.g., the co-existence of different residents in a city to enjoy urban open spaces), there is asymmetry (e.g., the benefits of urban open spaces more for the well-to-do but less for the homeless).

And in relation to the regression-progression principle, if there is progress (e.g., the construction of urban open spaces as an escape from life stress in urban density), there is regresssion (e.g., the subsequent problems associated with the construction of urban open spaces, such as urban sprawls, segmentation, class bias, etc.).

# Void and Society

A different way to examine urban void can be had from the perspective of society, in the context of sustainability in urban planning.

The perspective of society can be illustrated by way of two case studies, namely, (3.4.1) urban decay and its cycle, and (3.4.2) urbicide and urban warfare—to be addressed below, respectively.

## Urban Decay and Its Cycle

Any urban development in urban density does not last forever. Over time, it can degenerate into "urban decay," a kind of urban void in a city.

But exactly is this phenomenon of "urban decay"?

### *The Characteristics of Urban Decay*

The term "urban decay," in accordance to a conventional definition, refers to "a process by which a city, or a part of a city, falls into a state of disrepair." (WK 2009e)

There are many characteristics of urban decay, and hereafter are some of them, for illustration, as summarized in *Table 3.9*: (WK 2009e; R. Lupton 2004)

- "Depopulation"
- "Property abandonment"
- "Fragmented families"
- "High unemployment"
- "Political disenfranchisement"
- "Crimes"
- "Desolate and unfriendly urban landscapes"
- "Contaminated land and obsolete infrastructure"

Urban decay can occur differently in different cities. For instance, "in most cities in Europe and countries outside of North America," urban decay is "usually located on the outskirts of major metropolitan areas while the city center and inner city retain high real estate values and a steady or increasing population." (WK 2009e)

For instance, in Britain, urban decay occurred "in the 1970s and 1980s. Major cities like Glasgow in Scotland, the towns of the South Wales valleys, and some of the major English industrial cities like Birmingham, Manchester, Liverpool, Newcastle, and East London all experienced population decreases with very large areas of 19th century housing experiencing market price collapse." (WK 2009e)

In France, "large French cities are often surrounded by decayed areas. While the city center tends to be occupied mostly by middle- as well as upper-class residents, the city is often surrounded by very large mid to high-rise housing projects. The concentration of poverty and crime radiating from the developments often cause the entire suburb to fall into a state of urban decay as more affluent citizens seek housing in the city, or further out in semi-rural areas. In early November 2005, the decaying northern suburbs of Paris were the scene of severe riots sparked in part by the substandard living conditions in public housing projects." (WK 2009e)

By contrast, in North American cities, urban decay occurs in a different way, in that, in those cities suffering from urban decay, there is "an outflux of population to city suburbs or exurbs, as in the case of white flight," with the urban centers in decline. (WK 2009e)

A good example involves the city of Detroit, and its prosperity over the years has been greatly dependent on its vast "car manufacturing sector," which "employed the majority of its residents. When this industry began relocating outside of the city, it experienced massive population loss with associated urban decay, particularly after the 1967 riots. In 1950 the city's population was (according to the U.S. census) around 1.85 million; by 2003, this had declined to 911,000, a loss of nearly 940,000 people (52%)." (WK 2009e)

In these cities with urban decay, there are often "slums," where the residents suffer the most from urban decay. According to the United Nations agency, UN-HABITAT, a slum "is a run-down area of a city characterized by substandard housing and squalor and lacking in tenure security. The term has traditionally referred to housing areas that were once respectable but which deteriorated as the original dwellers moved on to newer and better parts of the city, but has come to include the vast informal settlements found in cities in the developing world." (WK 2009f; UNH 2007)

Many slums have some usual characteristics to be identified, although different slums in different countries may not share the same extent. Good examples of these characteristics are shown below: (WK 2009f)

- "High rates of poverty, and unemployment"
- "Breeding grounds" for "crime, drug addiction, alcoholism, high rates of mental illness, and suicide"
- "High rates of disease due to unsanitary conditions, malnutrition, and lack of basic health care"
- "Informal economy" (e.g., "street vending, drug dealing, domestic work, and prostitution")
- "Informal settlements," often with "poor structural quality of housing; overcrowding; and insecure residential status," which are vulnerable to "the brunt of natural and man-made disasters" like "landslides," "earthquakes," "tropical storms," and "fires")

Again, these characteristics are by no means exhaustive, but illustrative of what a slum in a city with urban decay is like.

*The Causes of Urban Decay*

An interesting question to ask of course is, What are the causes of urban decay? There is, however, no one single cause.

Here are some examples for thought, which can also be applied to "slums" in areas of urban decay, as summarized in *Table 3.9*: (WK 2009e & 2009h; R. Caro 1974; UNH 2007; W. Thabit 2005; P. Grogan 2001)

- "Urban planning mistakes" (e.g., messy housing projects for the poor)

- "Suburbanisation" (e.g., "white flight"), especially with "the development of freeways and railway lines"
- "Redlining" (viz., "the practice of denying or increasing the cost of services such as banking, insurance, access to jobs, access to health care, or even supermarkets to residents" of lower income and certain racial/ethnic groups
- "Rapid rural-to-urban migration"
- "Insecure [housing] tenure" (with "little incentive" for residents "to improve their surroundings"
- "Global economic booms and busts that ratchet up inequality and distribute new wealth unevenly"

It is one thing, of course, to identiy some major causes of urban decay, but the more vital question of resolving the chronic problem of urban decay is to be answered.

*Solutions to Urban Decay*

Urban planners over the years have proposed some solutions to fight urban decay. But how successful are they?

Let's consider two illustrative examples, namely, (a) gentrification and (b) planned shrinkage—to be addressed in what follows.

(a) Gentrification

The policy of "gentrification" (or "urban gentrification") is to encourage "the movement of more affluent individuals into a lower-class area," as a way to solve the problem of urban decay. (WK 2009i; B. Grant 2003)

It works something like this: "More households with higher incomes result in increased real estate values with higher associated rent, home prices, and property taxes. Industrial land use can decline with redevelopment bringing more commercial and residential use. Such changes often result in transformation of the neighborhood's character and culture." (WK 2009i; N. Brydson 2008)

This policy constitute an "urban reinvestment effort…by local governments or neighborhood groups, which directs money to invest in deteriorating city infrastructure, offer incentives for redevelopment, improve access to housing loans for low-income mortgage seekers, assist lending to first-time home purchasers, and improve rental properties. These efforts have been linked to reductions in local property crime rates, increased property prices, increased revenue to local governments from property taxes," and so on. (WK 2009i; A. Betsky 1997)

But for the critics, these benefits in gentrification actually contribute to the increase of urban decay in a different way, since it hurts "the neighborhood's lower-income residents. The increases in rent often result in the dispersal of communities whose members find that housing in the area is no longer affordable. Additionally, the increase in property taxes (due to increased property values) may sometimes force or give incentive for homeowners to sell their homes and move to less expensive neighborhoods. While those who view gentrification positively cite local reductions in a neighborhood's property crime rate, its critics argue that overall crime rates have not actually been reduced, but merely shifted to different lower-income neighborhoods." (WK 2009i; K. Pekarchik 2001; J. Lawless 2007; L. Reid 2003)

(b) Planned Shrinkage

If "gentrification" is not as successful as its proponents would like us to believe, there is another option, this time, in the form of "planned shrinkage."

The term "planned shrinkage" refers "a United States policy of withdrawing essential city services (such as police patrols, garbage removal, street repairs, and fire services) from neighborhoods suffering from urban decay, crime and poverty so that neighborhoods may be claimed by outside interests for new development." (WK 2009g)

The rationale of "planned shrinkage" is that, "after the mid-century boom in highways, suburbs and urban dispersal, civic leaders felt that urban decline was a natural and inevitable process, and they sought to plan the manner in which cities would shrink in such a way that population loss would be greatest in the areas with the poorest non-white residents." (WK 2009g; D. Wallace 2002; R. Gratz1989)

However, for the critics, the policy of "planned shrinkage" can easily backfire and also end up worsening urban decay for the poor, when the services withdrawn only make things worse for them.

The city of New York is a good example, in that, "by the mid 1970's, the Bronx had 120,000 fires per year (an average of about 3 fires every 2 hours); 40 percent of the housing in the area was destroyed....The overall rate of fires in the city shows a dramatic rise from about 60,000 reported fires in 1960 to rates of over 120,000 per year throughout most of the 1970's, peaking in 1980....The response time for fires also increased, as the firefighters simply did not have the resources to keep up. Many residents felt that the city was doing nothing to stop the fires." (WK 2009g)

In addition, "planned shrinkage also had impacts on public health. The pattern of the AIDS outbreak itself, has been gravely affected, and even strongly determined, by the outcomes of a program of planned shrinkage directed in African-American and Hispanic communities, and implemented through systematic and continuing denial of municipal services—particularly fire extinguishment resources--essential for maintaining urban levels of population density and ensuring community stability." (WK 2009g; R. Wallace 1990)

*The Cycle of Urban Decay*

This analysis of urban decay and the efforts to revive urban areas really reveals an urban matter of fact, in that urban decay can be "a natural and inevitable process," to the extent that no urban area can last forever, with the recurrent cycle of rise and fall. (WK 2009g)

This is true, even when any successful urban renewal efforts, even when they can produce improvement, will eventually suffer their own decline. So the cycle continues.

The implications for the issue concerning the possibility and desirability of sustainability in urban void should not be missed.

In relation to the issue of possibility, if urban density can over time lead to urban decay (as a form of urban void), the latter will also not last forever, with the ever recurrent cycle of rise and fall. Thus, the possibility of sustainability in urban void is hard to have, just as it is true for urban density.

And in relation to the issue of desirability, urban decay has its double sides, in that it can benefit some groups more than others; for instance, urban decay gives the opportunity for the dual efforts of gentrification and planned shrinkage, which benefit more the well-to-do than the less-well-to-do.

This is not to say, of course, that one should give up all efforts to fight urban decay, just as it is absurd to suggest that one should endorse any urban renewal wholeheartedly, without critical thinking. The point here is to remind us the double sides concerning the issue of possibility and desirability of sustainability, with the phenomenon of urban decay (as a form of urban void) as a case study here.

In the end, this insight can also reveal something else which is not less important, in showing us the relevance of the ontological principles in existential dialectics, and good examples include the absoluteness-relativeness principle, the denseness-emptiness principle, the simpleness-complicatedness principle, the change-constancy principle, the order-chaos principle, the expansion-contraction principle, the convention-novelty principle, the evolution-transformation principle, the symmetry-asymmetry principle, the regression-progression principle, and the same-difference principle.

For instance, in relation to the absoluteness-relativeness principle, if there is absolute viewpoint (e.g., the necessity to fight urban decay with planned shrinkage, gentrification, or something else), there is also relative standpoint (e.g., what is valid for the well-to-do in gentrification may not be so for the poor).

In relation to the denseness-emptiness principle, if there is emptiness (e.g., the existence of urban decay, say, in depopulation, as a form of urban void), there is density too (e.g., the efforts to revive central cities from urban decay to urban density).

In relation to the simpleness-complicatedness principle, if there is simplistic analysis (e.g., the simplistic view of planned shrinkage as a solution to urban decay), there is complicated concern (e.g., the more complicated analysis of planned shrinkage by taking into account the issues of economic inequality, the divide of public health, and the like).

In relation to the change-constancy principle, if there is change (e.g, the occurrence of urban decay and the attempt to solve it), there is constancy (e.g., the cycle of the rise and decline of urban cities).

In relation to the order-chaos principle, there is order (e.g., the planned efforts by urban planners to fight urban decay, as in the planned "movement of more affluent individuals into a lower-class area"), there is chaos too (e.g., the multifaceted uncertainties of living in urban decay for the poor).

In relation to the expansion-contraction principle, if there is expansion (e.g., the movement of the well-to-do in suburban areas, as in "white flight"), there is contraction (e.g., the subsequent effort for "planned shrinkage," as a way to fight urban decay).

In relation to the convention-novelty principle, if there is conventional way of thinking (e.g., the older way of suburbanization to escape from urban decay), there is novel challenge (e.g., the newer efforts in the form of gentrification and planned shrinkage, as an innovative way to fight urban decay).

In relation to the evolution-transformation principle, if there is evolution (e.g., the natural evoluton of humans to live in groups), there is transformation (e.g., the use of technology, as in the building of railroads and automobiles, for the suburban way of life).

In relation to the symmetry-asymmetry principle, if there is symmetry (e.g., the co-existence of urban decay and urban renewal), there is asymmetry (e.g., the worse impact of urban decay for the poor, and the better impact of urban renewal for the rich).

In relation to the regression-progression pricniple, if there is progress (e.g., the efforts for urban renewal by way of gentrification and planned shrinkage), there is regression (e.g., the actual worsening of urban decay in some other aspects of gentrification and planned shrinkage).

And in relation to the same-difference principle, if there is similarity in outcome (e.g., the privilege of the well-to-do to cope with urban renewal and urban decay), there is difference in outcome (e.g., suburbanization as an escape from urban decay, and gentrification as a way for urban renewal).

## Urbicide and Urban Warfare

Another way to examine urban void is to link urban warfare with urban design, to the extent that urban planners, architects, and the military work together to destroy enemy cities in the urban war on terror by terror, in what is now known as "urbicide" (for city destruction)—just as, by analogy, there is "homicide" (for the destruction, or killing, of an individual by another)

### *Urbicide and the Urban War on Terror by Terror*

Phillip Misselwitz (2003), in his conversation with Eyal Weizman (2003a), who, together with Rafi Segal, wrote *A Civilian Occupation: The Politics of Israeli Architecture*, introduced the contemporary phenomenon of "urbicide" (or "designed destruction"), as Steve Grahman put it, which has become "default global strategies in which the city is turned against its inhabitants," to the extent that "architects and planners face a minefield of new ethical dilemmas."

This is all the more urgent, especially (though not exclusively) since "the September 11th attacks, [when] the Western world is in the grip of the proclaimed 'War on Terrorism.'" (P. Misselwitz 2003)

The nature of "asymmetric warfare" becomes in vogue, as "the days of the Clausewitzian definition of warfare as a symmetrical engagement between state armies in the open field are over. War has entered the city again—the sphere of the everyday, the private realm of the house, sacred to Western societies. We find ourselves nervous when we use public transport systems or mingle in crowds, due to frequent bomb scares. Our parliaments are debating whether to grant powers to our armies to maintain internal security, powers that were previously held by the police, while violent clashes with the anti-globalisation movement take place on main public squares and shopping streets." (P. Misselwitz 2003)

For Eyal Weizman (2003), urban warfare is not new, since "cities were always exposed to war and organised according to the logic of defense. Each period's urban form related to the available technologies of destruction."

As cities grow over time, they become "security nightmares" for the military, and new ways to cope with them require different technologies and military tactics.

The reason is that "the military forces feel threatened by the huge increase of big cities and their ineptness at dealing with them. It is the very nature of urban areas and their tendencies to density, congestion, diversities, heterogeneity, and formal diversity that makes them hard to invade and conquer. The military tends to deal with the problem of taking over a city in a way similar to the way a planner deals with issues of development. Both look for ways to control an area by manipulating its infrastructure, reshaping and replacing the built fabric, or attempting to manage the local population's various cultural sensitivities." (E. Weizman 2003)

This history of urban warfare can be summarized by way of a few illustrations, namely, (a) the case of Algiers in 1840, (b) the case of Paris in the 1870's, (c) the case of Palestine in 1936, and (d) the case of Bethlehem in 2002—to be described in what follows.

(a) The Case of Algiers in 1840

A good starting point concerns the work by Sharon Rotbard, who described "the French 1840's invasion of Algiers by Marshall Thomas Robert Bugeaud. In a typically colonial attitude of zero tolerance and total disregard for the complexity of the historical structure, Bugeaud set out to break popular support for the resistance leader Abdel Kader by attacking the fabric of cities, towns, and villages. His actions were so extreme and brutal that they managed to raise parliamentary criticism in 19th century Paris." (E. Weizman 2003)

In spite of the criticism, Bugeaud succeeded in the end to destroy the city, with more than one hundred thousand troops," even though it took him "seven years to subdue Abdel Kader's ten-thousand man army. He finally regained control over Algiers' dense kasbahs by destroying entire neighbourhoods in reprisals for guerrilla attacks, sometimes breaking centres of resistance by reshaping cities, widening roads for military movements. These were some of the first demolitions used as military planning: Kader's resistance was broken," in the end. (E. Weizman 2003)

(b) The Case of Paris in the 1870's

Bugeaud, with the successful, though brutal, urban warfare in Algiers in mind, later wrote a book titled *La Guerre des Rues et des Maison* (or, in

English, *The War of Streets and Houses*) which "is described as the first manual for the preparation and conduct of urban warfare." (E. Weizman 2003)

The brutal French victory in Algiers also encouraged Bugeaud to use Algiers as "a laboratory for another war. The ardent Royalist Bugeaud personified the anti-urban attitudes of the French restoration. Opposed to the industrial revolution, which he thought was physically and morally toxic, he believed that he could reverse the trend of migration to cities by making land cultivation more efficient.," especially in relation to "the densely populated, desperately poor, and rapidly growing capital of Paris. As mass migration from rural France led to the dramatic overcrowding of cities' outer districts, new strategies to maintain stability and state control had to be developed." (E. Weizman 2003)

So, Bugeaud "proposed a radical reorganisation of the city [Paris]. Much like in Algiers,…new routes for military manoeuvres should be cut through the city and military regiments positioned within it. Bugeaud understood that there is a direct relationship between the organisation of the terrain and the ability to control it militarily and that this logic, when extended into the urban area, means that military thinking has to guide urban design. If strategic urban design previously focused on strengthening the city's peripheral walls and fortifications to keep out the enemy, here, since the enemy was already inside the city, the city had to be controlled from within. It is the city fabric itself, its streets and houses, that were to be adapted accordingly." (E. Weizman 2003)

With Bugeaud's proposal in mind, Georges-Eugène Haussmann led "the great modernisation project of Paris in the early 1870's….Thus it seems that the experiment of Algiers led, ironically, to one of the most influential and admired urban projects of the modern era. Haussmann created wide boulevards down which the cavalry could charge against rioting crowds and artillery would have a straight lines of fire to break barricades, while levelling many labyrinthine slums. Military control was exercised on the drawing board, according to the rules of design, fashion, and speculative interests. In the 19th century, one can perhaps still differentiate between the different kinds of destruction: while urban warfare is tested at the periphery of Western civilisation, the modernisation policies in the city centres employ the same tools, but camouflage themselves with different rhetoric." (E. Weizman 2003)

(c) The Case of Palestine in 1936

This legacy of "urban destruction" continued later in Palestine, with the euphonious name of "Operation Anchor," which was carried out by the

British to destroy the city of "Old Jaffa" in 1936, "a time later known as the first Arab rebellion, perhaps the first Intifada." (E. Weizman 2003)

The British forces initially "suffered casualties from stone throwers and snipers protected by Jaffa's winding roads and dense urban fabric. In response, the British government decided to cut a large anchor-shaped boulevard through the old city, destroying between 300 and 700 homes. This boulevard is still apparent today—it is where most of old Jaffa's cafes and restaurants are located, but at the time it allowed deep patrols into the very heart of the city and put an end to resistance with a combination of design and military force." (E. Weizman 2003)

But the British success, like the French one in Algiers, was inhumanely brutal, in terms of human and physical destruction. For instance, in Old Jaffa, "all complexities were flattened out, intellectually and physically. Attempts to 'understand' local cultures were distorted by a Romantic vision of the Mediterranean kasbah—a place that can be considered aesthetically fascinating, but that remains suspect, deceptive, treacherous, and violent....Knowledge and power are closely linked, and colonial knowledge was largely reproduced on maps." (E. Weizman 2003)

(d) The Case of Bethlehem in 2002

As cities grow larger and more complicated, "military personnel seek to learn how the cities work so that they may control them by manipulating their various components....Seizing control of cities is now perceived as the only effective way to control a state." (E. Weizman 2003)

Urban destruction in urban warfare now targets "all layers" of a city, such as "roads, power supplies, water, and communication networks can be more effective than controlling an urban space by conventional means. By temporarily shutting down electricity and telephone connections in particular parts of a city, the military can paralyse the enemy." (E. Weizman 2003)

In fact, "the most important precedent was set with the Israeli incursion into the refugee camps of the West Bank in April 2002. In view of the expected invasion of Baghdad, Jenin [the refugee camp complex] not only supplied a valuable source of information, but was also considered a live model for this emerging type of urban warfare. While governments and human rights organisations strongly condemned the acts of the Israeli army, militaries were eager for every piece of information provided by Israeli generals, through open and classified channels." (E. Weizman 2003)

The result was nothing but a brutal urbicide, although the battle at the beginning did not proceeded in the way that Israel would like: "In Jenin, a few dozen Palestinian fighters managed to hold back a whole army division as long as fighting took place between the homes and streets. The Israelis only 'won' the battle when bulldozers collapsed the city on its defenders. The complexities of urban warfare were finally erased in the last days of the battle, when the centre of the refugee camp, an area about 300 meters square, was flattened. 350 buildings, mostly homes, were destroyed or severely damaged, and about 4,000 people were left homeless." (E. Weizman 2003)

*Some Limits of Urbicide*

But urbicide has its limits, as "cities continue to reduce the advantages of the technologically superior force." (E. Weizman 2003)

Consider three illustrative limits below, namely, (a) urban complexity, (b) global media, and (c) international criminal court.

To avoid any false expectations, it should be stressed, however, that the limits are not very effective in stopping urbicide. In other words, urbicide can still occur, in spite of these limits.

(a) Urban Complexity

The first limit which reduces the effectiveness of urbicide concerns urban complexity. For instance, "the contemporary city has developed complexity, especially along the vertical axis. Its infrastructure—sewage systems, electric telecommunications, water mains, and underground transport systemis buried in the subsurface. Supersurfaces of very high roofscapes have emerged, while the air between and above them is cluttered with complex electromagnetic fields. Besides growing vertically, cities now sprawl horizontally across vast territorial regions. Within this type of environment, high-tech military equipment is easily incapacitated. Buildings mask targets or create urban canyons, which diminish the capabilities of the airforce. It is hard to see into the urban battle space; it is very difficult to communicate in it, because radio waves are often disturbed. It is hard to use precision weapons in it because it is difficult to obtain accurate GPS satellite locations." (E. Weizman 2003)

In light of this urban complexity, urbicide has to confront urban "chaos" and "unpredictability" in the process of urban destruction, "where all the plans and preparations became irrelevant," and the process of urbicide can be "completely unexpected, dense, full of contradictions, with characters changing their role from woman to man, from civilian to combatant, from friend to foe. Chance played a more important role than the ability to calculate and predict. It has become impossible to draw up scenarios, plan next steps, or draw up single-track plans to follow through." (E. Weizman 2003)

But this does not stop urbicide to continue. As shown in the case of Bethlehem in 2002 (as described in the previous section), the Israeli army succeeded in flattening out the entire city in a most brutal way, in spite of urban complexity.

(b) Global Media

The second limit concerns global media, with the instantaneous "reports on the atrocities committed by Western forces [which] have great effect on public opinion." (E. Weizman 2003)

Even then, however, urbicide can still occur, because "the media tends to collaborate with the military. Focusing on psychological warfare and public relations as a key element in urban warfare, the military has a clear interest in promoting maximum cooperation from the media and often uses it to disseminate information or disinformation and to maintain political support at home. In Iraq, we saw a lot of reporters attached to military units, essentially doing what military propaganda did generations before. It is revealing that the US military calls these methods a 'strategy of reprogramming mass consciousness.'" (E. Weizman 2003)

(c) International Criminal Court

The third limit involves international law against war crimes. This is true, especially after WWII, "we have seen the rise of international institutions and the elaboration of the laws of war. The inauguration of the ICC (International Criminal Court) in the Hague [in 2003]…made it possible to examine and prosecute individuals for war crimes." (E. Weizman 2003)

Again, urbicide can still occur, because the institution of international law has not been effective in stopping military violence by major powers.

The reason is that "the court's jurisdiction extends only to member states…. Since military planners are acutely aware that the methods required for urban warfare will make soldiers potentially liable to prosecution for war crimes, American and Israeli governments cancelled their memberships in the ICC." (E. Weizman 2003)

As Weizman (2003) thus critically observed, "as long as the world armies serve nations and are not under the authority of a single world government, the 'rule of law' between these nations is impossible. Some 'war criminals' may be taken to court while others not. The question of who will face trial will be dictated by power politics. But in the end, the mere knowledge that the ICC exists might help deter war crimes."

### *The Ethical and Legal Problems for Urban Planners*

The occurrence of urbicide, even with its limits, poses a serious dilemma to urban planners and architects, as they are often nowadays collaborating with the military for eliminating the others. So, a challenging question for them is, Are urban planners and architects liable to be held for war crimes?

As a start, the military is quire aware of the criminal consequences in urbicide and engages in a propaganda campaign to cover them up.

For instance, "American and Israeli military jargon calls these acts 'reshaping the battle space,' organising the city in a way that serves the attacking force….[And] the American General Keith Dickson defined campaigns of planned destruction as the 're-orientation of the built fabric to create conditions favourable for operational movement and manoeuvre.' It looks as if the military jargon is accustoming itself to a cleaner, publicly defendable language in which technical terminology is used to dress up the actions which include the levelling buildings to improve transportation, and the destruction of infrastructure to deny water, electricity, and other systems to the defenders." (E. Weizman 2003)

Even then, this military tendency to use euphonious words to cover up brutality and inhumanity in urbicide does not prevent some critics to raise the difficult question for urban planners and architects who collaborate with the military for designed destruction, as Phillip Misselwitz (2003) aptly asked, "If architects as experts of urban issues become willing informants and collaborators for the military—indirectly and directly—this action seems to touch upon fundamental ethical issues.."

Keeping in mind "the story of the crime of urbicide" in Bethlehem committed by Israeli army in 2002 (as told by Steve Graham)—Weizman (2003) thus answered the question posed by Misselwitz: "Architects and planners have the responsibility to use their ability to help make people understand the repercussions involved in formal aspects of warfare—and the crimes of an attack on urbanity. The large bulldozers employed by the Israeli army in the West Bank and Gaza to destroy homes were the most effective strategic urban weapon. Each one of these mammoths is manned by a crew of three, including an engineering officer—usually a civil engineer or an architect on reserve duty. The reason is that they best know how to topple a building, to which side the debris must fall, etc. This is similar to a medical doctor's engagement in torture. Architecture has no equivalent of the Hippocratic oath, but if we accept urbicide and destruction as war crimes, architects may in principle end up in jail."

*Urbicide and the Issue of Sustainability*

This analysis is to show us the extent of possibility and desirability of sustainability in urban planning when examined with the case study of urbicide in the context of urban void.

In relation to the issue of possibility, urbicide has its own limits, be they about urban complexity, international criminal court, and global media.

And in relation to the issue of desirability, urbicide raises some serious issues concerning moral and legal consequences, which may put urban planners and architects in jail or make them feel shameful.

Thus, the issue of sustainability, both in terms of possibility and desirability, is not as promising as urban planners would like us to believe, even though the case study here concerns urbicide in the context of urban void.

The anlaysis, in addition, reveals something interesting about the ontolological principles in existential dialectics, and good examples include the absoluteness-relativeness principle, the predictability-unpredictability principle, the denseness-emptiness principle, the simpleness-complicatedness principle, the openness-hiddenness principle, the order-chaos principle, the convention-novelty principle, the evolution-transformation princple, the symmetry-asymmetry principle, the softness-hardness principle, the regression-progression principle, and the same-difference principle.

For instance, in relation to the absoluteness-relativeness principle, if there is absolute thinking (e.g., the absolute viewpoint by the military to achieve urban victory, by any means necessary, like urbicide), there is relative viewpoint (e.g., what is valid for the military in urban warfare may not be so for the victims in urbicide).

In relation to the predictability-unpredictability principle, if there is predictability (e.g,., the predictable way of urban warfare by the military nowadays, like urbicide), there is unpredictability (e.g., the unpredictable nature of chance and complexity in the process of urbicide).

In relation to the denseness-emptiness principle, if there is denseness (e.g., the growth of large cities with huge population), there is emptiness (e.g., the need of urbicide in urban warfare by the military).

In relation to the simpleness-complicatedness principle, if there is simple analysis (e.g., the simplistic endorsement of urbicide by the military), there is complicated concern (e.g., the more complicated analysis of different "layers" of a city, be they horizontal or vertical, in urban warfare, together with ethical and legal issues).

In relation to the openness-hiddenness principle, if there is openness (e.g., the open campaign for national security in the war on terror), there is hiddenness (e.g., the cover-up of brutality and its criminal consequences, by way of euphonious usage of military jargons and the refusal to join ICC).

In relation to the order-chaos principle, if there is order (e.g., the planned destruction of a city in urban warfare), there is chaos (e.g., the chaotic process of urbicide when actually carried out).

In relation to the convention-novelty principle, if there is conventional wisdom (e.g., the conventional Clausewitzian definition of warfare), there is novel challenge (e.g., the novel invention of urbicide for urban warfare).

In relation to the evolution-transformation principle, if there is evolution (e.g., the violent conflicts in the state of nature, including those among primitive humans), there is transformation (e.g., the increasing use of technology in urban warfare, like precision-guided bombs or bulldozers).

In relation to the symmetry-asymmetry principle, if there is symmetry (e.g., the co-existence of conventional warfare and urban warfare), there is asymmetry (e.g., the popularity of conventional warfare among nation-states, but the relevance of urbicide in urban warfare).

In relation to the softness-hardness principle, if there is softness (e.g., the propagandistic use of euphonious words to win the heart and mind of the populace in a state of war), there is hardness (e.g., the use of urbicide to wipe out the entire city by brutal force).

In relation to the regression-progression principle, if there is regression (e.g., the brutal outcome of urbicide in urban warfare), there is progress (e.g., the better way to fight urban warfare by urbicide than by the conventional Clausewitzian means).

And in relation to the same-difference principle, if there is similarity in outcome (e.g., the destruction of properties and the environment, together with human casualties in war, be it conventional or urban), there is difference in outcome too (e.g., the destruction of the city in urbicide, as opposed to the destruction of the national territory in conventional warfare).

## Void and Culture

The last perspective to be considered here in the context of sustainability with urban void as a case study is about the role of culture.

Consider two illustrations, namely, (3.5.1) individualism and urban emptiness, and (3.5.2) the monastic emptiness, and urban suffering—to be analyzed below, in that order.

## Individualism and Urban Emptiness

Urban void, as an counterpoint to urban density, is the domain where the individual can seek refuge from the stress in urban life. When examined from the perspective of culture, urban void illuminates the nature of the yearning for the inner self in the context of urban emptiness.

The deeper question to ask, of course, is, To what extent is this search for the inner self possible and desirable as an escape from urban density, in the domain of urban void (from the perspective of culture).

*The Search for the Ideal Self*

Urban density comes along with its ever recurrent chronic problems which increase the level of urban stress, and good examples concern crowdedness, congestion, heterogeneity, and poverty at the social level—and, for that matter, conformist habits and practices at the cultural level.

In the current context of culture, urban density imposes certain habits and practices as in "mores" (viz., "the fixed morally binding customs") and "norms" (viz., the "principle[s] of right action binding upon the members of a group and serving to guide, control, or regulate proper and acceptable behavior." (WK 2009 & 2009a)

The cultural expectation of conformist habits and practices is not necessarily welcomed by anyone. Some individuals thus yearn for some breathing spaces in some degrees of urban void, where they can be at ease with their inner self with no higher authority to report to for approval and sanction.

Thus is the search for the self in urban void—away from the conformist habits and practices in urban density at the cultural level.

*The Attraction of Individualism*

Ralph Waldo Emerson, for one, sought for this inner self in urban void, away from the conformist habits and practices in urban density, when he asked us to "desert the tradition," since "the perpetual admonition of nature to us, is, 'The world is new, untried. Do not believe the past. I give you the universe a virgin today.'" (R. Bellah 1985)

This new universe that he sought in urban void was to place "new importance given to the single person," such that "everything…[is] to insulate the individual, to surround him with barriers of natural respect, so that each man shall feel the world is his, and man shall treat with man as a sovereign state with a sovereign state." (R. Bellah 1985; R. Emerson 1985)

He once thus recounted one of his inspiring moments when he was at ease with himself in the state of emptiness, away from urban density: "Crossing a bare common, in snow puddles, at twilight, under a clouded sky, without having in my thoughts any occurrence of special good fortune, I have enjoyed a perfect exhilaration. I am glad to the brink of fear...I become a transparent eyeball; I am nothing; I see all….The name of the nearest friend sounds then foreign and accidental: to be brothers, to be acquaintances, master or servant is then a trifle and a disturbance." (R. Bellah 1985)

For Robert Coles (1980), this search for the inner self "means a concentration persistent if not feverish upon one's thoughts feelings wishes, worries bordering on, if not embracing, solipsism: the self as the only or main form of (existential) reality."

In other words, the fanciful search for the inner self in individualism away from urban density is to idealize "the…self in its absolute freedom," which "is completely 'unencumbered.' to borrow a term from Michael Sandel [1982]….It presumes the existence of an abso1utely empty, unencumbered, and provisional self." (R. Bellah 1985)

The intention behind the search for the ideal self is sensible enough, since it strives to free oneself from the oppressive habits and practices in urban density, in the fear that "we tend to become interchangeable cogs in the social machinery, to become directionless and empty, to be led by the nose by whatever 'sells' in the marketplace, including a widespread 'personality market' in which even personal qualities must be revised to accommodate the impulses or preferences of others, and to treat others and ourselves as depersonalized objects." (K. Frost 2009)

As Erich Fromn (1975) put it, "we tend to sell out…to fanaticism, the illusion of total fulfillment in…craving and seeking the approval of others at all costs, numerous escapisms, or just going shopping." (K. Frost 2009)

Or, in "existentialist" jargon, the search for the ideal self in modern times is "born of a profound suspicion of social pressures toward conformity and inauthenticity in modern society and often have inspired people who feel trapped or dehumanized in an age given to conformism and scientific determinism. Partly for this reason, numerous modern therapy theories make a version of this ideal…central to their picture of human functioning." (K. Frost 2009; F. Richardson 1993)

This is all the more urgent, when the binding nature of the habits and practices in urban density has the notorious history of "religious wars, ethnic hatred, punishing discrimination, callous oppression, and the like." (K. Frost 2009)

But then, what is the price of this search for the ideal self in urban void, however justified though it may be?

*The Dark Side of the Individual Self*

For the critics, however, individualism like this is not sustainable, even if it comes with a noble intention.

To start, "the ideal self" so understood is just a fanciful notion, as it is nowhere to be found in real life. In fact, the ideal self so understood cannot sustain social relations which make life satisfactory for many.

For the communitarians, for instance, "only if one makes a personal, voluntary effort to 'get involved' in helping others will one's life be complete. And…many…do join voluntary associations, service clubs, and church societies whose main purpose is to help the unfortunate or better the community. Such activities bring great personal satisfaction. The happiness and joy of giving are earned by making a free individual decision to join such an organization, to accept its discipline, and to participate in its charitable work." (R. Bellah 1985)

For these individuals, "the self-interest demanded by the individualistic pursuit of success needs to be balanced by the sympathy of the voluntary act of concern for others. Without the joyful experience of support from such a community, an individual would find it difficult to make the effort to be a success, and success achieved would likely turn to ashes in his mouth." (R. Bellah 1985)

Yet, it is easier said than done. Those who argue for the individual self warn that "it is…no easy task to strike a balance between the kind of self-interest implicit in the individualistic search for success and the kind of concern required to gain the joys of community and public involvement." (R. Bellah 1985)

The reason of this inner tension is that, while "we strongly assert the value of our self-reliance and autonomy,…[w]e deeply feel the emptiness of a life without sustaining social commitments. Yet we are afraid to articulate our sense that we need each other as much as we need to stand alone for fear that if we did we would lose our independence altogether." (R. Bellah 1985)

This dilemma is all the more revealed in the lack of an "objective" moral order (like the habits and practices in urban density) that everyone can agree upon, without thinking that they are all grounded on power and self-interest.

The deeper problem here, in the end, is that, if the ideal self is nowhere to be found, neither is there the objective moral order for everyone. And this is precisely what I went to great lengths to explain in the two volumes of *The Future of Human Civilization* (*FHC*).

*A Daunting Task for Urban Planners*

But then, there is a daunting task awaiting urban planners, in how to sustain an urban design, to the extent that it can satisfy the needs of both the individualists and the communitarians, not that one is preferred over the other, since both of them are in conflict with each other.

This is easier said than done. And the fact that people from both sides of the fence have debated with (and, worse, fought against) each other since time immemorial speaks volumes of the deep problem here.

In the end, urban void is neither possible nor desirable to the extent that urban planners would like us to believe, for the reason as already explained above in regard to the different trade-offs experienced by both sides of the debate.

The analysis of this deep problem has the virtue to show us some ontological principles in existential dialectics, and good examples include the absolutness-relativeness principle, the partiality-totality principle, the denseness-emptiness principle, the simpleness-complicatedness principle, the convention-novelty principle, the symmetry-asymmetry principle, the softness-hardness principle, the regression-progression principle, and the same-difference principle.

For instance, in relation to the absoluteness-relativeness principle, if there is absolute viewpoint (e.g., the notorious absolute mentality in "religious wars, ethnic hatred, punishing discrimination, callous oppression, and the like"), there is likewise relative consideration (e.g., what is valid for the solitary in urban emptiness may not be so for the social in urban density).

In relation to the partiality-totality principle, if there is partial analysis (e.g., the narrow perspective of an individualist in relation to his own desires and values), there is totalistic counterpart (e.g., the larger picture of the whole society for an inclusive dialogue among different groups).

In relation to the denseness-emptiness principle, if there is emptiness (e.g., the search for the inner self within itself in urban emptiness), there is denseness (e.g., the search for the social self with others in urban density).

In relation to the simpleness-complicatedness principle, if there is simplistic analysis (e.g., the simplistic endorsement of deserting tradition in Emersonian individualism), there is complicated concern (e.g., the more complicated analysis of the problems associated with the "unencumbered" nature of the inner self).

In relation to the convention-novelty principle, if there is conventional wisdom (e.g., living with others in accordance to the habits and practices in culture), there is novel challenge (e.g., the Emersonian proposal of being "nothing" in a new universe without the tradition).

In relation to the symmetry-asymmetry principle, if there is symmetry (e.g., the co-existence of different individualist and communitarian views of personal autonomy in the context of modern urban life), there is asymmetry (e.g., the acceptance of individualism by individuals like Emerson, and the popularity of communitarianism by individuals like Sandel).

In relation to the softness-hardness principle, if there is softness (e.g., the individualist persuasion for a new universe by way of self-discovery), there is hardness (e.g., the violent history of "religious wars,...punishing discrimination, callous oppression, and the like" to force others to live in a certain way).

In relation to the regression-progression principle, if there is progress (e.g., the individualist rejection of the violent history of "religious wars, ethnic hatred, punishing discrimination, callous oppression, and the like" for a better lifeworld), there is regression (e.g., the feverish obsession, solipsism, and loneliness in the individualist self).

And in relation to the same-difference principle, if there is similiarity in outcome (e.g., the dilemma of trade-offs in each of the two different ways of life), there is difference in outcome (e.g., the different trade-off as experienced by either of the individualist and communitarian groups).

## The Monastic Emptiness, and Urban Suffering

If urban density can produce urban stress, which leads to the search for the inner self in urban void—there is another path, this time, by way of the practice of a monastic life to counter urban suffering.

A good case in point concerns the religion of Buddhism, which, "based on the teachings attributed to Siddhartha Gautama, commonly known as 'The Buddha' (the Awakened One),...shared his insights to help sentient beings end their suffering by understanding the true nature of phenomena, thereby escaping the cycle of suffering and rebirth (saṃsāra), that is, achieving Nirvana." (WK 2009k; R. Gethin 2008)

*The Idea of Karma*

A good starting point is to understand the concept of "Karma," which means "action, work" in Sanskrit and refers to "the energy which drives *Saṃsāra*, the cycle of suffering and rebirth for each being." (WK 20009k)

The idea of "Karma" is important in Buddhism, in that it is linked to those actions (of body, speech, and mind) that spring from mental intent (Pāli: *cetana*), and which bring about a consequence (or fruit, Sanskrit: *phala*) or result (Pāli: *vipāka*). Every time a person acts there is some quality of intention at the base of the mind and it is that quality rather than the outward appearance of the action that determines its effect." (WK 2009k)

All actions, be they "good" (or "kusala" in Pāli) or "bad" (or "akusala" in Pāli), "produce 'seeds' in the mind which come to fruition either in this life or in a subsequent rebirth. The avoidance of unwholesome actions and the cultivation of positive actions is called *Śīla* (from Sanskrit: ethical conduct)." (WK 2009k; T. Kasulis 2009)

*The Process of Rebirth*

These actions, be they good and bad, can be reborn in the cycle of life, death, and rebirth, albeit not exactly in the same way.

The idea of "rebirth" refers "to a process whereby beings go through a succession of lifetimes as one of many possible forms of sentient life, each running from conception to death. It is important to note, however, that Buddhism rejects concepts of a permanent self or an unchanging, eternal soul, as it is called in Christianity or even Hinduism. As there ultimately is no such thing as a self *(anatta)*, rebirth in subsequent existences must rather be understood as the continuation of a dynamic, ever-changing process of 'dependent arising' (*Pratītyasamutpāda*) determined by the laws of cause and effect *(Karma)* rather than that of one being, 'jumping' from one existence to the next." (WK 2009k; D. Keown 1996)

In Buddhism, rebirth can take place in different "realms" (which can even be subdivided further into different "planes of existence"), as shown below: (WK 2009k; S. Mahathera 2009)

- "Naraka beings: those who live in one of many Narakas (Hells)"
- "Animals: sharing some space with humans, but considered another type of life"
- "Preta: Sometimes sharing some space with humans, but invisible to most people; an important variety is the hungry ghost"

- "Human beings: one of the realms of rebirth in which attaining Nirvana is possible"
- "Asuras: variously translated as lowly deities, demons, titans, antigods; not recognized by Theravada (Mahavihara) tradition as a separate realm."
- "Devas including Brahmas: variously translated as gods, deities, spirits, angels, or left untranslated"

These different realms for rebirth, ranging from "hells" through "Nirvana" to "gods," can occur, dependent on the history of Karma in relation to the individual in the previous life.

*The Cycle of Samsara*

A major motivation for the Buddhists is to therefore strive for the ending of this cycle of suffering (or "the cycle of *Samsara*"), especially in the less desirable realms (e.g., "hells").

The underlyiing cause is that "sentient beings crave pleasure and are averse to pain from birth to death. In being controlled by these attitudes, they perpetuate the cycle of conditioned existence and suffering (*Samsara*), and produce the causes and conditions of the next rebirth after death. Each rebirth repeats this process in an involuntary cycle." (WK 2009k)

So, a good question to ask here is, But how is suffering to be ended?

*The Cessation of Suffering (Dukkha)*

A short answer is that suffering can be ended by freeing oneself from desire.

For instance, "according to the Pali Tipitaka, the Four Noble Truths were the first teaching of Gautama Buddha after attaining Nirvana," and the truths show us the way for this ascetic goal, as shown below: (WK 2009k; P. Thera 1999 )

- "Life as we know it ultimately is or leads to suffering (*dukkha*) in one way or another."
- "Suffering is caused by craving or attachments to worldly pleasures of all kinds. This is often expressed as a deluded clinging to a certain sense of existence, to selfhood, or to the things or people that we consider the cause of happiness or unhappiness."

- "Suffering ends when craving ends, when one is freed from desire. This is achieved by eliminating all delusion, thereby reaching a liberated state of Enlightenment (*bodhi*)."
- "Reaching this liberated state is achieved by following the path laid out by the Buddha."

And the path as laid out by the Buddha can be summarized in terms of "The Noble Eightfold Path," which can be classified in three categories, namley, for the purification of the mind, the ethics of abstention, and the mental discipline of the mind, as shown below: (WK 2009k)

- Prajñā: the "wisdom" to "purify" the mind
  —"dṛṣṭi (ditthi): viewing reality as it is, not just as it appears to be"
  —"saṃkalpa (sankappa): intention of renunciation,...and harmlessness"
- Śīla: "the morality, or abstention from unwholesome deeds"
  —"vāc (vāca): speaking in a truthful and non hurtful way"
  —"karman (kammanta): acting in a non harmful way"
  —"ājīvana (ājīva): a non harmful livelihood"
- Samādhi: the "mental discipline" over the mind
  —"vyāyāma (vāyāma): making an effort to improve"
  —"smṛti (sati)": "to see things...without any craving or aversion"
  —"samādhi (samādhi): correct meditation or concentration"

A common misunderstanding is to assume that the Buddhists are self-torturing to the extreme, with these ascetic commandments. But in Buddhism, there is the idea of "the Middle Way" to warn against any extreme self-torturing or self-mortification.

*The Middle Way*

The idea of "the Middle Way" can be understood in different senses, as shown below: (WK 2009k)

- "The practice of non-extremism: a path of moderation away from the extremes of self-indulgence and self-mortification"
- "The middle ground between certain metaphysical views (e.g., that things ultimately either do or do not exist)"
- "An explanation of Nirvana (perfect enlightenment), a state wherein it becomes clear that all dualities apparent in the world are delusory"

- "Another term for emptiness, the ultimate nature of all phenomena, lack of inherent existence, which avoids the extremes of permanence and nihilism or inherent existence and nothingness"

But this idea of "emptiness" (above) in Buddhism needs some clarifications, as given in the next subsection.

*The Monastic Emptiness*

The idea of "emptiness" in Buddhism is not the simple opposite of "something" but stands for the transcendence of all dualities in our everyday understanding of reality into the primordial oneness.

In the end, this transcendence is for the "cessation of suffering," and the monastic emptiness of existence is thus an existence which is free of "suffering" (or empty of suffering). (WK 2009m)

Buddhism offers diffferent explanations over time in regard to this transcendence into the primordial oneness. For illustration, here are three often-cited ones.

Firstly, the first explanation is that such search for dualities (e.g., "eternal and non-eternal," "the body and the self," "existence and inexistence") distracts us from "activity that is practical to realizing enlightenment and bring about the danger of substituting the experience of liberation by conceptual understanding of the doctrine or by religious faith." (WK 2009k; B. Thanissaro 1997; K. Werner 1989)

Secondly, another explanation is that "both affirmative and negative positions regarding these questions are based on attachment to and misunderstanding of the aggregates and senses. That is, when one sees these things for what they are, the idea of forming positions on such metaphysical questions simply does not occur to one." (WK 2009k; B. Thanissaro 2004)

And thirdly, the third explanation is that "reality is devoid of designations, or empty, and therefore language itself is a priori inadequate." (WK 2009k; G. Nagao 1991)

In so doing, the Buddhist approach focuses "on direct cognition of phenomena as they are presented to the mind. This informs and supports the Buddhist approach to liberation via ethical and meditative training known as the Noble Eightfold Path." (WK 2009k)

It is thus no wonder that, "in the Pali Canon and numerous Mahayana Sutras and Tantras, the Buddha is portrayed stressing that Dharma (in the sense of truth) cannot truly be understood with the ordinary rational mind or logic—reality transcends all worldly concepts." (WK 2009k)

And in "the Mahayana Mahaparinirvana Sutra's self-styled 'Uttara-Tantra,' the Buddha insists that…one must then relinquish fixation on words and letters, as these are utterly divorced from liberation….The Tantra entitled the 'All-Creating King' (Kunjed Gyalpo Tantra, recognized by the Nyingma school of Tibetan Buddhism) also emphasizes how Buddhist truth lies beyond the range of discursive/verbal thought and is ultimately mysterious. The Supreme Buddha, Samantabhadra, states there: 'The mind of perfect purity ... is beyond thinking and inexplicable.'" (WK 2009k; E. Neumaier-Dargyay 1992)

Only then is the door open to the primodial oneness, with "the permanent cessation of suffering" in the monastic emptiness. (WK 2009m)

*Buddhism in Reality*

All this talk of searching for enlightenment from urban suffering sounds very nice indeed, but to what extent does its implementation over the ages succed in reaching the ultimate goal of the monastic emptiness in "nirvana"?

Precisely here, there is no lack of criticisms against the Buddhist search for the monastic emptiness (as a form of urban void to combat suffering in urban density). Consider three major criticisms below, for illustration.

Firstly, the Buddhist monastic life is by no means so spiritual as claimed, because, more often than not, it can simply lead to "boredom."

Paul Breiter, for instance, spent some years practising Buddhism in South, East, and Southeast Asia and had this observation to make, when he was in a Burmese monastery: "Boredom, no doubt, accounts for the inordinate amount of sleeping one sees in monasteries—monks are forever taking naps—as well as for the dullness and apathy frequently encountered in them. I suspect too, that [they]…do so…as an escape from the tedium of monastic living. Similarly, boredom probably accounts for the great interest monks show in visitors." (S. Dhammika 2009)

Secondly, the Buddhist monastic life is not so ascetic as expected, because monks can easily become indulgent in expensive projects and lifestyles, without paying for them (or, at least, appreciating the hardships of others who have to endure to come up with the money for the expenses).

Breiter, for instance, when visiting Buddhist monasteries, were often struck with different "unrealistic, wasteful, ill-conceived or never completed religious monuments or projects….Why do monks so often involve themselves in such useless undertakings?...Many monks…can be assured of financial support from the devote and they will never be called upon to justify their proposal no matter how cockeyed or unnecessary it might be. No one would ever think to question a monk's judgment or obstruct his wishes….If his grandiose scheme falls through, goes bankrupt or is never finished he can just walk and leave others to deal with the mess….From the time he enters his monastery everything is provided for him on a silver platter, in some cases quite literally." (S. Dhammika 2009)

Moreover, "in places like the north and east of Sri Lanka and in up-country Burma there are monasteries that are relatively poor but even here the monks always live better than the lay people do. Most monasteries are well appointed, many, particularly in Thailand, could only be described as luxurious. In rural areas it is common that the local monastery is the only place in the village with a television, air conditioning or fans, cement buildings, tiled roof, running water, a car and servants." (S. Dhammika 2009)

H.L.Seneviratne thus offered an explanation about this paradox of their preaching ascetism and yet living an expensive life: "[T]he esteem monks are held in, their complete alienation from physical work and their comfortable lifestyles 'contributes to an explanation of the paradox of an allegedly infinitely compassionate order's appalling insensitivity to large-scale human suffering...The effect of these cultural notions also contributes to an isolation of monks from a realistic and felt idea of the economic hardship of the ordinary people.'" (S. Dhammika 2009)

Thirdly, the Buddhist monastic life is not so compassionate towards the others as assumed, because monks often practice exclusion.

Consider, say, the exclusion of women. As Paul Breiter reminded us, "monks treat all females as being physically and ritually impure. Thai monks will not take anything directly from a woman's hand and in Burma they will not visit the home of a woman who is menstruating nor will a woman in that condition visit a monastery or temple. In Thailand, particularly in the north, women are not allowed to circumambulate stupas because their inherent impurity will destroy the power of the relic within. In Burma they are not allowed to touch certain sacred Buddha images, enter simas or even some particularly holy shrines." (S. Dhammika 2009)

Besides, monks often engage in "sectarianism and ethnocentricity." For instance, "[i]n all Theravadin countries the Sangha is split into different sects—two in Thailand, three in Sri Lanka and more than half a dozen in Burma. In Sri Lanka the Siam and Amarapura Nikayas are further split into so-called chapters." (S. Dhammika 2009)

And, not only "there is…no unity between Buddhist monks of different countries," but also "there is actually hostility between them. The Burmese Bhikkhu has little regard for his Ceylon brother, and the latter regards the former with not as much affection as he should. The Chinese monk derides both as having 'incomplete views' and the Japanese has very scant knowledge of the Sangha in other countries." (B. Prajnananda 2009)

Even in the West, "Western Buddhists, and Western Buddhist institutions, are pretty overwhelmingly white and middle class and above, and don't seem to be particularly good or interested in reaching beyond those boundaries. Another issue is the isolation of 'ethnic' Buddhists in the West from converts." (WK 2009l)

*The Sustainability of Monastic Emptiness*

This critical analysis raises a more important question, in regard to the extent of possibility and desirability of the monastic emptiness as promised in the teaching of Buddhism, as a way to deal with urban suffering.

The promise of the monastic life on paper looks quite different, when it is applied in practice—even if the case study here is illustrative, not exhaustive. This then poses a further challenge to urban planners, in regard to how they can design a city which makes it easier for individuals to be free of urban suffering without the dark sides encountered in Buddhism in real life.

Yet, the analysis is by no means useless, since it can illuminate some ontological principles in existential dialectics, and good examples include the absoluteness-relativeness principle, the formalness-informalness principle, the denseness-emptiness principle, the simpleness-complicatedness principle, the openness-hiddenness principle, the theory-praxis principle, the symmetry-asymmetry principle, the softness-hardness princple, and the regression-progression principle.

For instance, in relation to the absolutness-relativeness principle, if there is absolute viewpoint (e.g., the individualist craving for the selfish pleasures of the flesh in urban life), there is relative counterpart (e.g, the Buddhist view of the relative interdependence of all selves, so what is valid for the individualist may not be so for the Buddhist).

In relation to the formalness-informalness principle, if there is formal logic (e.g., the law of the excluded middle in treating the pleasure and pain in an either-or dichtomy, as in Western individulalism), there is informal one too (e.g., the alternative logic, say, among white Buddhists, in treating pleasure and pain as a dialectic duality, such that pleasure and pain can occur together).

In relation to the denseness-emptiness principle, if there is denseness (e.g., the concentration of indivdual selves in urban density), there is emptiness (e.g., the yearning for monastic emptiness, away from the suffering in urban density).

In relation to the simpleness-complicatedness principle, if there is simple thinking (e.g., the simplistic understanding among lay people of the ascetic life in Buddhist monasteries as a solution to urban suffering), there is complicated analysis too (e.g., the more complicated analysis of the ascetic life in its mundane details, when applied in practice).

In relation to the openness-hiddenness principle, if there is openness (e.g., the open preaching of the virtue of monastic life), there is hiddenness (e.g., the taboo or reluctance among monks to tell others about the boredom, extravagence, and exclusiveness in the monastic world).

In relation to the theory-praxis princple, if there is theoretical construction (e.g., the Buddhist idea of the monastic life), there is its practical application (e.g., the application of Buddhist teaching in monasteries).

In relation to the symmetry-asymmetry principle, if there is symmetry (e.g., the co-existence of different sects or demoninations of Buddhism), there is asymmetry (e.g., the popularity of Zen Buddhism in Japan, but the acceptance of Theravada Buddhism in Southeast Asia).

In relation to the softness-hardness principle, if there is softness (e.g., the teaching of compassion and non-violence in Buddhism), there is hardness (e.g., the exclusiveness and violence among different sects and denominations of Buddhism).

And in relation to the regression-progression principle, if there is regression (e.g., the dark sides of Buddhism in practice, be they about boredom, exclusiveness, and indulgence), there is progress (e.g., the Buddhist contribution to the freedom from urban suffering by way of the monastic life in urban void).

## The Repulsion of Void

At this juncture, it is therefore pertinent to ask, To what extent can sustainbility (in the context of urban void as a case study here) be achieved? And to what extent can sustainability be desirable?

As illustrated from the four different perspectives of the mind, nature, society, and culture in this chapter, the overall picture no longer looks as rosy as the proponents of urban void would like us to believe.

In this light, if urban void has its attractiveness, it likewise has its repulsiveness, for reasons as already given in this chapter.

Consequently, the daunting challenge to urban planners is to design a city which makes it easier for individuals to strive for urban void without much of its dark sides.

In any event, the next important question to ask of course is, But what conclusion can then be drawn, at the end of this rather tedious journey of examining the nature of double sides of both urban density (in Chapter Two) and urban void (in this chapter), even if they serve as a case study for the issue concerning the possibility and desirability of sustainability?

This is the last question which is to be answered in the next chapter, to which we now turn.

**Table 3.1. Void and Its Solitude**

- **Void and the Mind**
  —Ex: Indifference and Urban Solitude
  —Ex: Urban Homelessness and Mental Health

- **Void and Nature**
  —Ex: Deforestation and Urban Destructiveness
  —Ex: Urban Open Space and Its Controversy

- **Void and Society**
  —Ex: Urban Decay and Its Cycle
  —Ex: Urbicide and Urban Warfare

- **Void and Culture**
  —Ex: Individualism and Urban Emptiness
  —Ex: The Monastic Emptiness, and Urban Suffering

*Notes*: The examples in the categories are solely illustrative (not exhaustive), and the comparison is relative (not absolute), nor are they necessarily mutually exclusive. And some can be easily re-classified elsewhere. As generalities, they allow exceptions.
*Source*: A summary of Ch.3 of *FPHUP*

**Table 3.2. Face-To-Face and Side-by-Side Relations (Part I)**

- **Face-to-Face Relations**

—For Fran Tonkiss (2003), in "face-to-face" relations, the community in question can take different social imaginations, of course, especially for "particular groups—often defined ethnically or spatially—as constituencies or objects of government, tracing a boundary around what are then produced as common interests or problems." (F. Tonkiss 2003)

—In this language of community, "face-to-face" relations can provide the basis "for mobilization, for opposition, for a positioning of and claim to voice. Here, discourses of community frame an assertive rather than a defensive politics of difference, a collective agent in arguments for social and cultural recognition and the pressing of group or minority rights." (F. Tonkiss 2003; S. Benhabib 1996; A. Gutmann 1994; C. Taylor 1992). In this sense, "face-to-face" relations represent the idea of "recognition," in the parlance of Denise Riley (2002), for those who are psychologically predisposed to be "gregarious," in that "[i]t assumes a certain sociality, a visible presence in social space." (F. Tonkiss 2003)

*(continued on next page)*

**Table 3.2. Face-To-Face and Side-by-Side Relations (Part II)**

- **Side-by-Side Relations**
  —By contrast, in "side-by-side" relations, the individuals in question psychologically prefer "to be unseen, to be unexceptional, to be impersonal in a social field where 'differences remain unassimilated'...and strangeness a matter of fact." (F. Tonkiss 2003; I. Young 1990)
  —In this sense, "side-by-side" relations constitutes the desire of "loneliness and privacy," to the extent that "cities" are now regarded by these individuals as "places in which relations of non-identity are possible, tolerable, even normal....This is not simply to set indifference in the city against the various workings of community or identity, the solitude of the street versus the convivial politics of the street party; rather, it is to suggest that part of a politics of difference...is when differences go unremarked because unremarkable, where otherness is ordinary, where a logic of anonymity displaces one of visibility....An 'ethics' of indifference may be (and perhaps usually is) inadvertent or unthinking, and the licence of anonymity, 'the right to be lonely' granted by the lack of concern of others, may be a temporary, uneven and unreliable freedom." (F. Tonkiss 2003)

*Sources*: From *Sec. 3.2.1* of *FPHUP*. For details, see F. Tonkiss (2003) and I. Young (1990) in the book.

**Table 3.3. Some Basic Characteristics of the Urban Homeless (Part I)**

- Marital Status: "72.3 percent…were single, 8.7 percent…were separated, 14.3 percent…were divorced, and 0.7 percent…were widowed. This speaks to the social isolation experienced by the individuals in the study sample."
- Education: "64.3 percent of the study sample did not complete high school."
- Income: "37.7 percent of the study sample have no income" at all.
- Encounters with the Legal System: "Between 50-60 percent of the study sample had encounters with the legal system at some point in their lives."
- Nights Spent in Police Stations or Jails in Year Prior to Interview: "People spent a lot more time in police stations or jails (30 percent ) than they did in psychiatric facilities (6 percent )."
- History of Homelessness: "The study sample showed 42 percent experiencing their first episode of homelessness, 44 percent having a pattern of chronic homelessness (three or more episodes), and 14 percent who had been homeless twice."
- Age of First Episode of Homelessness: "Homelessness is a pattern that develops relatively early in life: 131 subjects were homeless before the age of 18 years; by 30 years of age, 2/3 of the sample had experienced their first episode."
- Nights Spent on Street in Year Prior to Interview: "Over 60 percent of the sample spent time on the street, with two thirds spending more than 2 weeks on the street."
- Health Problems and Mental Illness: "People in the study sample reported chronic health problems, handicaps, disabilities, injuries and short-term illnesses. These were primarily orthopedic, respiratory and circulatory problems. While a very small proportion of the sample reported tuberculosis, it is known that tuberculosis is becoming a very significant and serious health problem in the homeless population."
- History of Childhood Abuse: "The rate of childhood sexual abuse and physical abuse reported by women and men in the homeless population was high compared with rates reported in the general population. Childhood sexual abuse: women (48.5 percent vs. 12.8 percent ); men (16 percent vs. 4 percent). Physical abuse: women (51 percent vs. 21 percent ); men (38 percent vs. 31.2 percent )."

*(continued on next page)*

**Table 3.3. Some Basic Characteristics of the Urban Homeless (Part II)**

- Substance Abuse and Mental Illness: The use of "alcohol, marijuana and cocaine in particular) is a major problem in this population. When prevalence rates of substance abuse (drug and alcohol abuse) are added to prevalence rates of mental illness, the overall prevalence rate rises to 86 percent . Only 14 percent of the study sample had no diagnosis of either mental illness or substance abuse. The prevalence rate of alcohol and substance abuse is almost identical to the prevalence rate of mental illness; roughly two thirds of the entire homeless population. This is 4-5 times the prevalence rate in the general population. Almost everyone with a lifetime diagnosis of mental illness also had a diagnosis of substance abuse. Three quarters of the people in every diagnostic category of mental illness also had substance abuse disorders."

*Notes*: The examples in the categories are solely illustrative (not exhaustive), and the comparison is relative (not absolute), nor are they necessarily mutually exclusive. And some can be easily re-classified elsewhere. As generalities, they allow exceptions.
*Sources*: From *Sec. 3.2.2* of *FPHUP*. For details, see D. Wasylenki (1998).

**Table 3.4. Causes and Sufferings of Urban Homelessness**

- **Causes of Urban Homelessness**
  —Ex: "Loss of job or having inadequate income to pay rent"
  —Ex: "Interpersonal conflict and divorce"
  —Ex: "Eviction"
  —Ex: "Drug and/or alcohol abuse"
  —Ex: "Mental illness"

- **Sufferings of Urban Homelessness**
  —Ex: "Legal problems"
  —Ex: "Illness of a loved one"
  —Ex: "Break up of a relationship"
  —Ex: "Debt"
  —Ex: "Physical abuse"
  —Ex: "Robbed"
  —Ex: "Reduction of benefits"
  —Ex: "Death of a loved one"

*Notes*: The examples in the categories are solely illustrative (not exhaustive), and the comparison is relative (not absolute), nor are they necessarily mutually exclusive. And some can be easily re-classified elsewhere. As generalities, they allow exceptions.
*Sources*: From *Sec. 3.2.2* of *FPHUP*. For details, see P. Goering (1998).

**Table 3.5. Causes of Anthropogenic Deforestation**

- Ex: "Farming" (e.g., "slash-and-burn")
- Ex: "Cattle ranching" (a form of animal husbandry)
- Ex: "Wood extraction" ("logging")
- Ex: "Infrastructure expansion"
- Ex: "Mining"
- Ex: "Population growth"
- Ex: "Urbanization"

*Notes*: The examples in the categories are solely illustrative (not exhaustive), and the comparison is relative (not absolute), nor are they necessarily mutually exclusive. And some can be easily re-classified elsewhere. As generalities, they allow exceptions.
*Sources*: From *Sec. 3.3.1* of *FPHUP*. For details, see WK (2009), H. Geist (2002), and SD (2008) in the book.

**Table 3.6. Impacts of Anthropogenic Deforestation (Part I)**

- **Atmospheric**
  —Anthropogenic deforestation "is a contributor to global warming, and is often cited as one of the major causes of the enhanced greenhouse effect." (WK 2009)
  —As an illustration, "tropical deforestation is responsible for approximately 20% of world greenhouse gas emissions. According to the Intergovernmental Panel on Climate Change, deforestation, mainly in tropical areas, account for up to one-third of total anthropogenic carbon dioxide emissions." (WK 2009)

- **Hydrologic**
  —For instance, "trees extract groundwater through their roots and release it into the atmosphere. When part of a forest is removed, the trees no longer evaporate away this water, resulting in a much drier climate." (WK 2009)
  —Also, "shrinking forest cover lessens the landscape's capacity to intercept, retain and transpire precipitation. Instead of trapping precipitation, which then percolates to groundwater systems, deforested areas become sources of surface water runoff, which moves much faster than subsurface flows. That quicker transport of surface water can translate into flash flooding and more localized floods than would occur with the forest cover." (WK 2009)

*(continued on next page)*

**Table 3.6. Impacts of Anthropogenic Deforestation (Part II)**

- **Soil**
  —In the state of nature, "undisturbed forest has very low rates of soil loss, approximately 2 metric tons per square kilometre (6 short tons per square mile)," but "deforestation generally increases rates of soil erosion, by increasing the amount of runoff and reducing the protection of the soil from tree litter….Forestry operations themselves also increase erosion through the development of roads and the use of mechanized equipment." (WK 2009)
  —A classic example concerns the Yellow River of China, in that "China's Loess Plateau was cleared of forest millennia ago. Since then it has been eroding, creating dramatic incised valleys, and providing the sediment that gives the Yellow River its yellow color and that causes the flooding of the river in the lower reaches (hence the river's nickname 'China's sorrow')." (WK 2009)

- **Ecological**
  —As an illustration, "the removal or destruction of areas of forest cover has resulted in a degraded environment with reduced biodiversity. Forests support biodiversity, providing habitat for wildlife; moreover, forests foster medicinal conservation. With forest biotopes being irreplaceable source of new drugs (such as taxol), deforestation can destroy genetic variations (such as crop resistance) irretrievably." (WK 2009; J. Stock 2009; SD 2009b; B. Fruth 2009; BM 2007)
  —This is all the more serious in certain areas of the earth, like the tropical rainforests, which "are the most diverse ecosystems on earth," and "about 80% of the world's known biodiversity could be found in tropical rainforests," just as any "removal or destruction of significant areas of forest cover has resulted in a degraded environment with reduced biodiversity." (WK 2009; BBC 2009 & 2009a; TQ 2009; M. Mogato 2008; R. Butler 2007)

*(continued on next page)*

**Table 3.6. Impacts of Anthropogenic Deforestation (Part III)**

- **Economic**
  —For instance, a report at the Convention on Biological Diversity (CBD) meeting in Bonn, Germany showed that the "[d]amage to forests and other aspects of nature could halve living standards for the world's poor and reduce global GDP by about 7% by 2050….Today, developed countries continue to utilize timber for building houses, and wood pulp for paper. In developing countries almost three billion people rely on wood for heating and cooking." (WK 2009; AAAS 2009; R. Black 2008)
  —More often than not, "[s]hort-term economic gains made by conversion of forest to agriculture, or over-exploitation of wood products, typically leads to loss of long-term income and long term biological productivity (hence reduction in nature's services). West Africa, Madagascar, Southeast Asia and many other regions have experienced lower revenue because of declining timber harvests. Illegal logging causes billions of dollars of losses to national economies annually." (WK 2009; R. Butler 2007)

*Notes*: The examples in the categories are solely illustrative (not exhaustive), and the comparison is relative (not absolute), nor are they necessarily mutually exclusive. And some can be easily re-classified elsewhere. As generalities, they allow exceptions.
*Sources*: From *Sec. 3.3.1* of *FPHUP*. For details, see WK (2009), NASA (2004), R. Mweninguwe (2005), SD (2009, 2009a & 2009b), C. Matthews (2009), P. Fearnside1 (2004), FC (2008), K. Denman (2007), I. Prentice (2009), J. Stock (2009), B. Fruth (2009), BM (2007), BBC (2009 & 2009a), TQ (2009), M. Mogato (2008), R. Butler (2007), AAAS (2009), and R. Black (2008) in the book.

**Table 3.7. Main Functions of Urban Open Spaces (Part I)**

- **Recreational**
  Urban open space is important, in that it can provide recreation as a way to reduce life stress in urban density. There are two main types of recreation in urban open space, in that it can be "active" (e.g., "organized sports and individual exercise") or "passive" (e.g., "being in the open space"). (WK 2009a) In this way, "[t]ime spent in an urban open space for recreation offers a reprieve from the urban environment." (WK 2009a)

- **Ecological**
  The second function of urban open space is "the conservation of nature in an urban environment." (WK 2009a) This focus on ecology in urban open space is to enhance our awareness "of the balance of nature, of natural processes and of man's place in and affect on nature—i.e., 'ecological awareness'....As human live more and more in man-made surroundings—i.e., cities—he risks harming himself by building and acting in ignorance of natural processes." (WK 2009a) In addition, "[b]eyond this man-nature benefit, urban open spaces also serve as islands of nature, promoting biodiversity and providing a home for natural species in environments that are otherwise uninhabitable due to city development." (WK 2009a)

- **Aesthetic**
  The third function is to create "aesthetic value," which means that "[p]eople enjoy viewing nature, especially when it is otherwise extensively deprived, as is the case in urban environments. Therefore, open space offers the value of 'substituting gray infrastructure.'" (WK 2009a; M. Eysenbach 2008)

*(continued on next page)*

**Table 3.7. Main Functions of Urban Open Spaces (Part II)**

- **Medical**
  And the fourth function of urban open space is to "offer health benefits to city residents through exposure to a natural environment." (WK 2009a) For instance, Richard Louv (2006) in *Last Child in the Woods* told us "the positive effects nature in general has on children, even when experienced in small islands of green within a city. Urban open spaces offer citizens relief from the strains of urban environments and everyday demands. That respite can come in the form of a walk or run, time spent sitting or reading, watching the birds, essentially any time spent in the natural environment the open space offers." (WK 2009a)

*Notes*: The examples in the categories are solely illustrative (not exhaustive), and the comparison is relative (not absolute), nor are they necessarily mutually exclusive. And some can be easily re-classified elsewhere. As generalities, they allow exceptions.
*Sources*: From *Sec. 3.3.2* of *FPHUP*. For details, see WK (2009a), L. Springgate (2008), BMR (1971), M. Eysenbach (2008), D. Berry (1976), and R. Louv (2006) in the book.

**Table 3.8. Main Problems of Sustainability in Urban Open Spaces (Part I)**

- **In Relation to the Possibility of Sustainability**

—Firstly, not all species can survive in an urban open space. For instance, "species which are most often able to co-exist with man in an urban setting are usually those which 'are able to reproduce rapidly and to take advantage of transitory conditions or to evolve varieties suited to the urban situation.' Therefore, larger urban open spaces, especially those with various types of environments, are more likely to support a diverse ecosystem." (WK 2009a; D. Gill 1973) But then, this requires a lot of resources (which are often not available) to fund the construction of "large urban open spaces which rely on a natural local ecosystem…in terms of maintaining a balanced biodiversity." (WK 2009a) In addition, larger urban open spaces often pose another challenge, in that, to be sustainble, they must be properly "established and managed primarily to benefit natural wildlife populations in order that they may function as regional reservoirs." (WK 2009a; D. Gill 1973) It is easier said than done, however, epsecially when conflicting interests among different social groups render general agreement difficult to achieve.

—Secondly, the increase of urban population over time often competes with urban open spaces (which are not unlimited) for different uses, especially when the problems involve "overcrowding, poverty, squalor, ill-health, lack of morals and morale, and so on." (WK 2009a)

—And thirdly, urban open spaces can evolve over time into different segments, such that it is difficult to integrate them all in a sustainable urban environment.

*(continued on next page)*

**Table 3.8. Main Problems of Sustainability in Urban Open Spaces (Part II)**

- **In Relation to the Desirability of Sustainability**
  —Firstly, urban open spaces, however well-intentioned as the urban planners are in designing them, often end up preserving "the middle class status quo" and benefiting its hidden interests. (WK 2009a)
  —Secondly, urban open spaces can often constitute serious barriers to urban expansion, which then lead to urban sprawl, with subsequent problems which are undesirable and, for that matter, unsustainable too. In other words, in the current example of greenbelt, "it might spur the growth of areas much further away from the city core than if it had not existed, thereby actually increasing urban sprawl." (WK 2009c)
  —And thirdly, urban open spaces can evolve over time into something else altogether, which was not what the original urban planning was for. For instance, "there are many examples whereby the actual effect of green belts is to act as a land reserve for future freeways and other highways. Examples include sections of the 407 highway north of Toronto and the Hunt Club Rd /Richmond Rd. south of Ottawa. Whether they are originally planned as such, or the result of a newer administration taking advantage of land that was left available by its predecessors, is debatable." (WK 2009c)

*Notes*: The examples in the categories are solely illustrative (not exhaustive), and the comparison is relative (not absolute), nor are they necessarily mutually exclusive. And some can be easily re-classified elsewhere. As generalities, they allow exceptions.
*Sources*: From *Sec. 3.3.2* of *FPHUP*. For details, see WK (2009a & 2009c), D. Gill (1973), and others in the book.

**Table 3.9. Characteristics and Causes of Urban Decay**

- **Characteristics of Urban Decay**
  —Ex: "Depopulation"
  —Ex: "Property abandonment"
  —Ex: "Fragmented families"
  —Ex: "High unemployment"
  —Ex: "Political disenfranchisement"
  —Ex: "Crimes"
  —Ex: "Desolate and unfriendly urban landscapes"
  —Ex: "Contaminated land and obsolete infrastructure"

- **Causes of Urban Decay**
  —Ex: "Urban planning mistakes" (e.g., messy housing projects for the poor)
  —Ex: "Suburbanisation" (e.g., "white flight"), especially with "the development of freeways and railway lines"
  —Ex: "Redlining" (viz., "the practice of denying or increasing the cost of services such as banking, insurance, access to jobs, access to health care, or even supermarkets to residents" of lower income and certain racial/ethnic groups
  —Ex: "Rapid rural-to-urban migration"
  —Ex: "Insecure [housing] tenure" (with "little incentive" for residents "to improve their surroundings"
  —Ex: "Global economic booms and busts that ratchet up inequality and distribute new wealth unevenly"

*Notes*: The examples in the categories are solely illustrative (not exhaustive), and the comparison is relative (not absolute), nor are they necessarily mutually exclusive. And some can be easily re-classified elsewhere. As generalities, they allow exceptions.
*Sources*: From *Sec. 3.4.1* of *FPHUP*. For details, see WK (2009e & 2009h), R. Lupton (2004), R. Caro (1974), UNH (2007), W. Thabit (2005), and P. Grogan (2001) in the book.

# • PART FOUR •

## *Conclusion*

# CHAPTER 4
# *CONCLUSION—THE FUTURE OF URBAN PLANNING*

What we do today largely depends on how we interpret the past, and our interpretation of the past will, to a considerable extent, be shaped by the futures that our desires have already created.
—M. Schwartz & M. Thompson (L. Hempel 1996)

## The Concern with Sustainability

The analysis of urban density (in Chapter Two) and urban void (in Chapter Three) has the intellectual virtue to illustrate the very extent of the possibility and desirability of sustainability, within the context of urban planning (as a case study).

Contrary to the contemporary wisdom in our times, the very realization of sustainability in urban planning, in a small sense—and in relation to other highly related yet very serious challenges in our times (e.g., ecological, economic, demographic, technological, moral, spiritual, and the like), in a broad sense—presupposes some conditions about the human mind, nature, culture, and society, which, when closely examined, are neither possible nor desirable to the extent that its proponents would so often like to claim.

The very discourse on sustainability constitutes another intellectual fad, which, like other fads before it, so often comes and goes in history. After all, nothing lasts foreever anyway, neither does the fashionable discourse on sustainability. Each urban design for a new city comes, with the welcome of a new era, just as it fades away, to the replaced by a new one in a different era later.

Thus continues the cycle of change—but with its constancy of recurrence.

In this way, the major theoretical approaches in the existing literature on urban planning (or sustainability in a broader sense) are inadequate, for reasons as already explained in previous chapters.

My original contribution to the debate is a contingent analysis of urban planning (or sustainability in a broader sense), which takes into account the four perspectives of the mind, nature, society, and culture. In this light, the word "contingent" in the title of the theory is suggestive, as it takes into account the contingent perspectives of the mind, nature, society, and culture.

In other words, this book provides an alternative (better) way to understand the nature of sustainability in urban planning, which learns from different sides of the debate but in the end transcends them all.

More specificially, *my contingent theory of urban planning* contains seven major theses, namely, (4.2) the first thesis: the denseness-emptiness principle, (4.3) the second thesis: the absoluteness-relativeness principle, (4.4) the third thesis: the simpleness-relativeness principle, (4.5) the fourth thesis: the evolution-transformation principle, (4.6) the fifth thesis: the symmetry-asymmetry principle, (4.7) the sixth thesis: the regression-progression principle, and (4.8) the seventh thesis: the post-human planning—to be summarized hereafter, respectively.

In this way, the contingent theory of urban planning is useful in a three-fold way, (a) to reveal something important about the ontological principles of existential dialectics at the meta-theoretical level, (b) to enhance the better understanding of the subject matter of urban planning (and sustainability, in a broader sense), and (c) to provide a vision of the future of urban planning to come that we have never known, as will be clear later.

## Other Relevant Principles

Lest misunderstanding hastily occurs (as this is something that I regularly did in my previous books, *almost verbatim*), it should be clarified at the outset that, if other ontological principles in existential dialectics (other than the six as listed in the seven theses) are not explicitly included in the listing of the seven theses, it is not because they are not important, but merely because they are not as equally relevant as the other six as indicated for the present study.

This is true, even if different studies of the same kind can yield different views about the degree of relevance for each ontological principle, depending on the specific nature of a research in question, needless to say. So, if a different author analyzes the subject matter in a different way, the relevance of the ontological principles will be different.

After all, the symmetry-asymmetry principle already implies that not all principles are equally relevant in any topic under examination, as they are not created equal for all cases under all circumstances (which would be an extreme view, of course).

In any event, the relevancy of other principles (besides the six as already listed in the theory above) for the present study can be summarized below, in relation to the five categories of ontological principles in existential dialectics.

*In the Category of Method*

Firstly, in regard to the formalness-informalness principle (on the formal requirements of logical systems), if there formal logic (e.g., the requirement of consistency in the rationalist mode of logic in *Sec. 2.5.1*; and the law of the excluded middle in treating the pleasure and pain in an either-or dichtomy, as in Western individulalism, in *Sec. 2.5.2*), there is likewise informal logic (e.g., the allowance of inconsistency and contradiction in yin-yang philosophy in *Sec. 2.5.1*; and the alternative logic, say, among white Buddhists, in treating pleasure and pain as a dialectic duality, such that pleasure and pain can occur together, in *Sec. 2.5.2*).

Secondly, in regard to the partiality-totality principle (on the relationships between whole and parts), if there are partial viewpoints (e.g., the narrow focus by each individual for his own needs in using natural resources in *Sec. 2.3*; the individual focus on his own self-interest in deciding whether to pollute or not in *Sec. 2.4.1*; the focus on the individual factors for someone to decide whether or not to adopt an AFV in *Sec. 2.4.3*; the engagement in logging from the individual self-interest of a wood-extraction company in *Sec. 3.3.1*; the local design of each urban open space in *Sec. 3.3.2*; and the narrow perspective of an individualist in relation to his own desires and values in *Sec. 3.5.1*), there is also totalistic ones too (e.g., the bigger picture of ecolological interdependence connecting all individuals in their relation to nature in *Sec. 2.3*; the tragedy of the commons when viewed from the larger picture in *Sec. 2.4.1*; the concern with the larger, uncertain outcomes of the system as a whole in the AFV market in *Sec. 2.4.3*; the tragedy of the commons, as in deforestation, when looked from the larger perspective of nature and society, in *Sec. 3.3.1*; the emergence of "segmentation" over time from the overall picture, as more urban open spaces are built, in *Sec. 3.3.2*; and the larger picture of the whole society for an inclusive dialogue among different groups in *Sec. 3.5.1*).

Thirdly, in regard to the predictability-unpredictability principle (on the occurrence of events), if there is predictability (e.g., the concern with sustainability when the tipping point of violating the balance with nature is crossed in *Sec. 2.3*; the prediction of more availability of AFVs over time in *Sec. 2.4.3*; the higher chance to become homeless for individuals who suffer from mental illness in *Sec. 3.2.2*; the prediction of soil erosion, with the increase of deforestation, in *Sec. 3.3.1*; the predictable benefits of building urban open spaces in *Sec. 3.3.2*; and the predictable way of urban warfare by the military nowadays, like urbicide, in *Sec. 3.4.2*), there is unpredictability too (e.g.,. the difficulty to know in advance which specific ways to collect and use LEK from the participants in a given community in *Sec. 2.3*; the inability to tell in advance which AFV markets will succeed and which ones will not in *Sec. 2.4.3*; the inability to predict urban homeless on the basis of mental illness alone, since not every homeless individual is mentally sick, in *Sec. 3.2.2*; the inability to make accurate "predictions about the impact of deforestation on biodiversity" in *Sec. 3.3.1*; different unintented problems over time to haunt any well-intentioned design of urban open spaces in *Sec. 3.3.2*; and the unpredictable nature of chance and complexity in the process of urbicide in *Sec. 3.4.2*).

And fourthly, in regard to the explicability-inexplicability principle (on the underlying mechanicisms of things), if there is explicability (e.g., a decrease of road capacity can lead to the increase of traffic density in *Sec. 2.2.1*; mental illness as a cause of urban homelessness in *Sec. 3.2.2*; and cattle ranching as a cause of anthropogenic deforestation in *Sec. 3.3.1*), there is inexplicability (e.g., the unsettled issue concerning the bio-psychological causes of road rage in *Sec. 2.2.1*; the inexplicability of why some individuals who are not mentally ill but become homeless, if mental illness is solely considered in the explanation, in *Sec. 3.2.2*; and the lack of general agreement on the specific nature of the causes of deforestation, with different scholars offering different viewpoints, in *Sec. 3.3.1*).

*In the Category of Structure*

Firstly, in regard to the preciseness-vagueness principle (on the refinement of things), if there is vagueness (e.g., the conceptualization of only three general models of LEK in sustainabile urban planning in *Sec. 2.3*; and the vague generalization of sound preferences for all in *Sec. 2.5.2*), there is also the need for preciseness (e.g., the further subdivision of M2a and M2b in the context of LEK in *Sec. 2.3*; and the more precise distinction between "sound field" and "soundscape" for the finer individual experience of space in *Sec. 2.5.2*).

Secondly, in regard to the open-hiddenness princple (on the detection of things), if there is openness (e.g., the exploration of the serious problem of pollution for the public in *Sec. 2.4.1*; the open exploration of different patterns for city planning in *Sec. 2.5.1*; the exploration of how to build sustainble community in urban density in *Sec. 3.2.1*; the exploration of the causes of urban homelessness *in Sec. 3.2.2*; the exploration of the functions of urban open spaces for all residents in a city in *Sec. 3.3.2*; the open campaign for national security in the war on terror in *Sec. 3.4.2*; and the open preaching of the virtue of monastic life in *Sec. 3.5.2*), there is hiddenness (e.g., the hidden interests of politicians, regulators, and interest groups, in light of the weak monitoring by the public, in *Sec. 2.4.1*; the religious dogma behind the search for the "holy city" in Sec. 2.51; "the admission of even occasional loneliness remains taboo," in D. Riley's words, in *Sec. 3.2.1*; the reluctance by many researchers, due to the fear of political incorrectness, to emphasize mental illness in the analysis of urban homelessness, in *Sec. 3.2.2*; the hidden class bias in the construction of urban open spaces in *Sec. 3.3.2*; the cover-up of brutality and its criminal consequences, by way of euphonious usage of military jargons and the refusal to join ICC, in *Sec. 3.4.2*; and the taboo or relunctance among monks to tell others about the boredom, extravagence, and exclusiveness in the monastic world in *Sec. 3.5.2*).

And thirdly, in regard to the finiteness-transfiniteness principle (on the nature of numbers), if there is finiteness (e.g., the multipe actors in the collusion among regulators, legislators, and interest groups in *Sec. 2.4.1*; and the different individuals in urban density who can relate to the urban solitude of Kathy Nguyen in *Sec. 3.2.1*), there is also transfiniteness (e.g., the millions, or more than millions, of voters or residents in a metropolis who are weak in monitoring government effectiveness, in *Sec. 2.4.1*; and the even greater number, maybe in millions or billions, of those residents in metropolis around the world trying to search for a community suitable to them in urban density, in *Sec. 3.2.1*).

*In the Category of Process*

Firstly, in regard to the change-constancy principle (on the alteration of things), if there is change (e.g., the ever new evolution of "economic shift, technological advances, and public expectations" in relation to urban density, in *Sec. 2.2.2*; and the occurrence of urban decay and the attempt to solve it in *Sec. 3.4.1*), there is constancy (e.g., the persistent existence of social cohesion and social breakdown over time in *Sec. 2.2.2*; and the cycle of the rise and decline of urban cities in *Sec. 3.4.1*).

Secondly, in relation to the order-chaos principle (on the pattern of things), if there is order (e.g., the construction of roads and the formulation of traffic regulations for orderly traffic in *Sec. 2.2.1*; the patterns of ecological interdependence in nature in *Sec. 2.3*; the creation of order in a city planning in *Sec. 2.5.1*; the planned efforts by urban planners to fight urban decay, as in the planned "movement of more affluent individuals into a lower-class area," in *Sec. 3.4.1*; and the planned destruction of a city in urban warfare in *Sec. 3.4.2*), there is also chaos (e.g., the occurrence of traffic jam and accidents in *Sec. 2.2.1*; the uncertain outcomes in a complex system in *Sec. 2.3*; the occurrence of chaos within a city, especially in what the Christians called the "evil city," in *Sec. 2.5.1*; the multifaceted uncertainties of living in urban decay for the poor in *Sec. 3.4.1*; and the chaotic process of urbicide when actually carried out in *Sec. 3.4.2*).

And thirdly, in regard to the slowness-quickness principle (on the speed of change), if there is quickness (e.g., faster traffic in roads with greater capacity in *Sec. 2.2.1*; and the ability of other individuals who are able to bounce back from experiencing life stresses in urban life and later become wiser and more successful in life in *Sec. 3.2.2*), there is slowness too (e.g., the occurrence of slower traffic in cases of traffic congestion in *Sec. 2.2.1*; and the slow recovery of some individuals after experiencing life stresses in urban life and later becoming homeless in *Sec. 3.2.2*).

And fourthly, in regard to the expansion-contraction principle (on the growth of things), if there is expansion (e.g., the expansion of an urban area into its suburban space in *Sec. 2.2.2*; the occurrence of urban sprawl in *Sec. 2.4.2*; the expansion of urban areas into more and more previously forested ones in *Sec. 3.3.1*; the construction of larger urban open spaces in *Sec. 3.3.2*; and the movement of the well-to-do in suburban areas, as in "white flight" in *Sec. 3.4.1*), there is contraction (e.g., the occurrence of urban decay in the aftermath of urban flight in *Sec. 2.2.2*; the problem of urban decay over time in Sec. 2.4.2; the increase of deforestation which renders some desertified areas unsuitable for human habitation in *Sec. 3.3.1*; the barriers to urban expansion with the construction of more urban open spaces in *Sec. 3.3.2*; and the subsequent effort for "planned shrinkage," as a way to fight urban decay, in *Sec. 3.4.1*).

*In the Category of Agency*

Firstly, in regard to the theory-praxis principle (on the duality of knowledge), if there is theoretical construction (e.g., the mousetrap theory of sustainable technologies in *Sec. 2.4.3*; and the Buddhist idea of the monastic life in *Sec. 3.5.2*), there is also its practical application (e.g., the practical manufacturing of AFVs by different car firms over time, be they successful or not, in *Sec. 2.4.3*; and the application of Buddhist teaching in monasteries in *Sec. 3.5.2*).

Secondly, in regard to the convention-novelty principle (on the nature of creative thinking), if there is conventional wisdom (e.g, the traditional "polluter pays" solution in *Sec. 2.4.1*; the old-fashioned advocation of urban renewal in *Sec. 2.4.2*; the popular talk of AFVs to fight the problem of climate change in *Sec. 2.4.3*; the older way of suburbanization to escape from urban decay in *Sec. 3.4.1*; the conventional Clausewitzian definition of warfare in *Sec. 3.4.2*; and living with others in accordance to the habits and practices in culture in *Sec. 3.5.1*), there is likewise novel challenge (e.g., the original Coase's theorem to fight pollution in *Sec. 2.4.1*; the original replacement of "urban renewal" with "community development" in *Sec. 2.4.2*; the original view by Sterman on how bad things can happen to sustainable technologies in *Sec. 2.4.3*; the newer efforts in the form of gentrification and planned shrinkage, as an innovative way to fight urban decay in *Sec. 3.4.1*; the novel invention of urbicide for urban warfare in *Sec. 3.4.2*; and the Emersonian proposal of being "nothing" in a new universe without the tradition in *Sec. 3.5.1*).

And thirdly, in regard to the softness-hardness principle (on the force of change), if there is softness (e.g., the legal enforcement against road rage as a way to promote public safety and traffic civility in *Sec. 2.2.1*; the nice civility to live together in social cohesion in certain areas of urban density in *Sec. 2.2.2*; the propagandistic use of euphonious words to win the heart and mind of the populace in a state of war in *Sec. 3.4.2*; the individualist persuasion for a new universe by way of self-discovery in *Sec. 3.5.1*; and the teaching of compassion and non-violence in Buddhism in *Sec. 3.5.2*; ), there is hardness too (e.g., the aggressive form of driving in road rage at the expense of other vehicles in *Sec. 2.2.1*; the occurrence of violent, criminal behaviors by some individuals against others in urban areas in *Sec. 2.2.2*; the use of urbicide to wipe out the entire city by brutal force in *Sec. 3.4.2*; the violent history of "religious wars,…punishing discrimination, callous oppression, and the like" to force others to live in a certain way in *Sec. 3.5.1*; and the exclusiveness and violence among different sects and denominations of Buddhism in *Sec. 3.5.2*).

*In the Category of Outcome*

In relation to the same-difference principle (on the metamorphosis of change), if there is difference in outcome (e.g., the lessening of urban pollution over time in some cases in *Sec. 2.4.1*; the ever new variations of city formation over time, be they about the "holy city" or the "evil city" in *Sec. 2.5.1*; suburbanization as an escape from urban decay, and gentrification as a way for urban renewal, in *Sec. 3.4.1*; the destruction of the city in urbicide, as opposed to the destruction of the national territory in conventional warfare, in *Sec. 3.4.2*; and the different trade-off as experienced by either of the individualist and communitarian groups in *Sec. 3.5.1*), there is similarity in outcome too (e.g., the persistent existence of collusion among different groups with their ulterior interests in *Sec. 2.4.1*; the recurrent drive for the "great city," be it in the image of Man or that of God, in *Sec. 2.5.1*; the privilege of the well-to-do to cope with urban renewal and urban decay in *Sec. 3.4.1*; the destruction of properties and the environment, together with human casualties in war, be it conventional or urban, in *Sec. 3.4.2*; and the dilemma of trade-offs in each of the two different ways of life in *Sec. 3.5.1*).

# 1st Thesis: The Denseness-Emptiness Principle

With these other principles in mind—the first thesis can be called, in the absence of better words, the denseness-emptiness principle (on the distribution of entities in space) in the category of "structure" in existential dialectics, which is first proposed in *FPHUP*, that is, in this book, in that both density and void are needed, in relation to the mind, nature, culture, and society, albeit in different ways.

Lest any misunderstanding occurs, the term "void" is used here only as an approximation of emptiness (depending on the degree of the lack of density), since, in physics, it is well known that "empty" space is not really empty all the way, because it can be full of energy (e.g., random quantum fluctuations at the sub-atomic level, and, for that matter, dark energy in the universe) and matter (e.g., different versions of sub-atomic particles, and, for that matter, dark matter in the universe). (F. Wilczek, Frank 2008)

With this qualification in mind, hereafter are some instances of density and void, as used in existential dialectics.

For instance, on the one hand, there is density (e.g., the requirement of urban density for large-scale infrastructure to be built in *Sec. 2.2.1*; the need of urban density for a large population in *Sec. 2.2.2*; the greater use of resources as the number of population in a habitat increases in *Sec. 2.3*; the occurrence of pollution, be it noise, air, or something else, in a densely populated environment, in *Sec. 2.4.1*; the policy of urban renewal to cope with the problem of urban density in *Sec .2.4.2*; the highly croweded urban environment, with its ever increasing demand for energy resources, like gasoline consumption, in *Sec. 2.4.3*; the problem of crowdedness in a dense environment, such that city planning is needed, even often tainted with the religious surge for order and purity, in *Sec. 2.5.1*; the existence of urban density in the context of sound pollution in *Sec. 2.5.2*; the existence of "face-to-face" relations in urban density in *Sec. 3.2.1*; the formation of close community for urban residents in *Sec. 3.2.2*; the increase of urban density, and its subsequent demand for more food production in *Sec. 3.3.1*; urban expansion with the increase of urban population in *Sec. 3.3.2*; the existence of urban decay, say, in depopulation, as a form of urban void, in *Sec. 3.4.1*; the growth of large cities with huge population in *Sec. 3.4.2*; the search for the inner self within itself in urban emptiness in *Sec. 3.5.1*; and the concentration of indivdual selves in urban density in *Sec. 3.5.2*).

On the other hand, there is void (e.g., the occurrence of road rage as a personal attempt to fight traffic density for a less croweded traffic space in *Sec. 2.2.1*; the need of less dense urban space for urban flight in *Sec. 2.2.2*; the need to restore the balance with nature by making a habitat less dense, say, by way of "low-intensity habitation," in *Sec. 2.3*; the search for an escape from urban density, such as fighting pollution or seeking a greener open space, in *Sec .2.4.1*; the subsequent urge for less dense environment by way of urban sprawl and later suburban sprawl, in the aftermath of urban renewal, in *Sec. 2.4.2*; the search for smaller AFVs and greener space, to minimize urban crowededness in *Sec. 2.4.3*; the yearning for a garden of paradise without the problems associated with urban density in *Sec. 2.5.1*; the search for "quiet space" in an urban setting in *Sec. 2.5.2*; the existence of "side-by-side" relations in urban void in *Sec. 3.2.1*; the wandering of the homeless in the street without home nor community in *Sec. 3.2.2*; the occurrence of deforestation and its subsequent desertification and other consequences, which make the affected areas unhabitable, in *Sec. 3.3.1*; the need for urban open spaces as an escape from life stress in urban density in *Sec. 3.3.2*; the efforts to revive central cities from urban decay to urban density in *Sec. 3.4.1*; the need of urbicide in urban warfare by the military in *Sec. 3.4.2*; the search for the social self with others in urban density in *Sec. 3.5.1*; and the yearning for monastic emptiness, away from the suffering in urban density, in *Sec. 3.5.2*).

It should be stressed, however, as I often stressed this in my previous books (*almost verbatim*), that there are different shades of gray (or different degrees of truth) in the two opposites in any principle, and the classification is not necessarily mutually exclusive either. In fact, even when some combinations of the two occur, they only end up sharing the same dialectic relationship, but in a different degree.

Also, there are some other relationships (as a kind of family resemblance) which have something in common with the principle, but they are not exactly the same but only more or less comparable, which varies from case to case.

Good examples of family resemblance in relation to the principle are fullness-void, and the like.

## 2nd Thesis: The Absoluteness-Relativeness Principle

The second thesis refers to the absoluteness-relativeness principle (on the multiplicity of things) in the category of "method" in existential dialectics, which was first proposed in *FPHK*, in that there is the multiplicity of things in reality, be they about entities, qualities (or properties), and relationships. And, if there is something absolute, there is likewise something relative.

Both absoluteness and relativeness here are also relevant to different modalities often cited in the literature on ontology, such as possibility (e.g., something "can" happen) and its opposite (e.g., impossibility), probability (e.g., something "will" happen) and its opposite (e.g., improbability), and necessity (e.g., something "should" happen) and its opposite (e.g., contingency).

For instance, on the one hand, there is absolute viewpoint (e.g., the focus on the good side of human nature for moral intergenerational equity, that is, to act now for the future persistent existence of posterity, in *Sec. 2.2.1*; the sole concern with the positive aspects of urban density in *Sec. 2.2.2*; the sole concern of greedy individuals with using natural resources for their own needs in *Sec. 2.3*; Coase's obsession with the free market to solve the pollution problem in *Sec. 2.4.1*; the euphoric advocation of urban renewal, especially by Robert Moses in 1930's, in *Sec. 2.4.2*; the one-dimensional focus on the building of sustainable technologies to fight the problem of climate change in *Sec. 2.4.3*; the absolute viewpoint in religious fundamentalism in *Sec. 2.5.1*; the search for universal sound preferences in *Sec. 2.5.2*; the fashionable talk of building community in urban density in *Sec. 3.2.1*; the blaming by some social workers on culture and society for for the problem of urban homelessness, as revealed in the Workshop of the Mental Health Policy Research Group, where some argued that urban homelessness is a "social," not "individual," problem, in *Sec. 3.2.2*; the euphoric discourse on urban open spaces in *Sec. 3.3.2*; the necessity to fight urban decay with planned shrinkage, gentrification, or something else, in *Sec. 3.4.1*; the absolute viewpoint by the military to achieve urban victory, by any means necessary, like urbicide, in *Sec. 3.4.2*; the notorious absolute mentality in "religious wars, ethnic hatred, punishing discrimination, callous oppression, and the like," in *Sec. 3.5.1*; and the individualist craving for the selfish pleasures of the flesh in urban life in *Sec. 3.5.2*).

On the other hand, there is likewise relative concern (e.g., the more balanced analysis of both good and dark sides of human nature, such as the issues concerning the short-term, biased, and shallow thinking process and the non-neutrality of human emotions, so that what is valid from one standpoint may not so from another, in *Sec. 2.2.1*; the twin aspects of social cohesion and social breakdown in urban density, such that what is valid from one standpoint may not be so from another, in *Sec. 2.2.2*; the dual concerns with using and exploiting natural resources in the relative viewpoint of balancing with nature, so that what is valid from one standpoint may not be so from another, in *Sec. 2.3*; the more balanced analysis of the pros and cons of both the market and government solutions, so that what is valid from one standpoint may not be so from another, in *Sec. 2.4.1*; the more balanced analysis of both benefits for the rich and the costs for the poor and the minorities in urban renewal, so that what is valid from one standpoint may not be so from another, in *Sec. 2.4.2*; the more balanced linkage between the invention of technologies and the wider social and cultural factors, so that what is valid from one standpoint may not be so from another, in *Sec. 2.4.3*; the allowance of uncertainty in chaos theory, so that what is valid, or certain, from one standpoint may not be so from another, in *Sec. 2.5.1*; the relative experience of space as shown in different soundscapes by each listener from the same acoustic source, so that what is pleasant from one standpoint may not be so from another, in *Sec. 2.5.2*; the yearning for solitude in urban void from the Others, so that what is valid from one standpoint may not be so from another, in *Sec. 3.2.1*; the response by Wasylenki, to show that the extreme view of blaming society and culture is too absolute, as both social and individual factors are important for the study of urban homelessness, such that things are more relative, or what is valid from one standpoint may not be so from another, in *Sec. 3.2.2*; there is relative counterpart (e.g., the relative analysis of class bias in urban open spaces, such that what is valid for the middle class may not be so for the homeless, in *Sec. 3.3.2*; what is valid for the well-to-do in gentrification may not be so for the poor in *Sec. 3.4.1*; what is valid for the military in urban warfare may not be so for the victims in urbicide in *Sec. 3.4.2*; what is valid for the solitary in urban emptiness may not be so for the social in urban density in *Sec. 3.5.1*; and the Buddhist view of the relative interdependence of all selves, so what is valid for the individualist may not be so for the Buddhist, in *Sec. 3.5.2*).

Again, it should be stressed, however, that there are different shades of gray (or different degrees of truth) in the two opposites, and the classification is not necessarily mutually exclusive either. In fact, even when some combinations of the two occur, they only end up sharing the same dialectic relationship, but in a different degree.

Also, there are some other relationships (as a kind of family resemblance) which have something in common with the principle, but they are not exactly the same but only more or less comparable, which varies from case to case.

Good examples of family resemblance in relation to the principle are uniformity-diversity, and so forth.

# 3rd Thesis: The Simpleness-Complicatedness Principle

The third thesis is called the simpleness-complicatedness principle (on the interconnection among things)) in the category of "structure" in existential dialectics (which was first out worked out in *FIA*), in that both simpleness and complicatedness are vital, without favoring one over the other, and each is utilized, depending on the basis of the perspectives of nature, the mind, culture, and society.

For instance, on the one hand, there is simple analysis (e.g., the simplistic analysis of the correlation between traffic density and road rage in *Sec. 2.2.1*; the call for sustainable urban planning with the use of LEK in *Sec. 2.3*; Coase's naïve faith on the free market to fight pollution in *Sec. 2.4.1*; the single-minded focus on restructuring a vast section of a city for commercial, government, and social services in traditional urban planning, in *Sec. 2.4.2*; the advocation of AFVs by its proponents, with little detailed analysis of the wider challenge of sustainability, in *Sec. 2.4.3*; the simplistic focus on the "technical" concerns with noise pollution in *Sec. 2.5.2*; the death of Kathy Nguyen due to anthrax by inhalation in *Sec. 3.2.1*; the simplistic focus on substance abuse and mental illness in addressing the characteristics of urban homelessness in *Sec. 3.2.2*; the account for the individual causes of deforestation in *Sec. 3.3.1*; the simple illustration of four functions of urban open spaces in *Sec. 3.3.2*; the simplistic view of planned shrinkage as a solution to urban decay in *Sec. 3.4.1*; the simplistic endorsement of urbicide by the military in *Sec. 3.4.2*; the simplistic endorsement of deserting tradition in Emersonian individualism in *Sec. 3.5.1*; and the simplistic understanding among lay people of the ascetic life in Buddhist monasteries as a solution to urban suffering in *Sec. 3.5.2*).

On the other hand, there is also complicated counterpart (e.g., the deeper issue concerning the bio-psychological causes of road rage in *Sec. 2.2.1*; the complicated analysis of different serious difficulties in collecting and using LEK in *Sec. 2.3*; a more complicated analysis of multiple actors and factors in the process of fighting pollution, as shown in the market and government models, in *Sec. 2.4.1*; the more complicated consideration of the plight of the ecological environment, the social structure, and community development in any urban planning, in *Sec. 2.4.2*; the more complicated account for AFVs, in relation to the multiple possibilities between the quality of technological innovation and its use or adoption, in *Sec. 2.4.3*; the more complicated analysis of how to fight noise pollution to include technical, aesthetic, social, and political problems, in *Sec. 2.5.2*; the more complicated, detailed analysis of Nguyen's death in the context of urban solitude in *Sec. 3.2.1*; the more complicated analysis of other main characteristics of urban homelessness, such as education, income, marital status, history of homelessness, childhood abuse, and the like, in *Sec. 3.2.2*; the need to look into the more complicated interrelations among the different causes of deforestation in *Sec. 3.3.1*; alternative but more complicated classifications of the functions of urban open spaces, to include, say, "economic," "cultural," "civic," "contemplative," etc., in *Sec. 3.3.2*; the more complicated analysis of planned shrinkage by taking into account the issues of economic inequality, the divide of public health, and the like in *Sec. 3.4.1*; the more complicated analysis of different "layers" of a city, be they horizontal or vertical, in urban warfare, together with ethical and legal issues, in *Sec. 3.4.2*; the more complicated analysis of the problems associated with the "unencumbered" nature of the inner self in *Sec. 3.5.1*; and the more complicated analysis of the ascetic life in its mundane details, when applied in practice, in *Sec. 3.5.2*).

Once more, it should be stressed, however, that there are different shades of gray (or different degrees of truth) in the two opposites, and the classification is not necessarily mutually exclusive either. In fact, even when some combinations of the two occur, they only end up sharing the same dialectic relationship, but in a different degree.

Also, there are some other relationships (as a kind of family resemblance) which have something in common with the principle, but they are not exactly the same but only more or less comparable, which varies from case to case.

Good examples of family resemblance in relation to the principle are inflexibility-flexibility, standardization-specialization, and the like.

## 4th Thesis: The Evolution-Transformation Principle

The fourth thesis concerns the evolution-transformation principle (on the multiple kinds of agency) in the category of "agency" in existential dialectics, which was first worked out in *FAE*, in that both the evolution in the state of nature (e.g., an object of natural beauty) and the transformation in the world of intelligent lifeforms (e.g., a work of art, an air-conditioner, etc.) are important.

And the transformative part of the principle gives technology a major role to play in the change of the world, which is something that I extensively analyzed in *FHC* in the context of the technophilic lifeworld, espeically though not exclusively since modern times.

For instance, on the one hand, there is natural evolution (e.g., the physical ability of humans to walk or run in *Sec. 2.2.1*; the natural evolution of humans to move around in *Sec. 2.2.2*; the natural evolution of humans to live in relation to nature in *Sec. 2.3*; the natural evolution of humans to walk uprght and run in *Sec. 2.4.3*; the natural evoution of humans to hear sounds from natural objects in *Sec. 2.5.2*; the natural causes of deforestation in *Sec. 3.3.1*; the natural evolution of humans to live in caves with nature, in much earlier times of history, in *Sec. 3.3.2*; the natural evoluton of humans to live in groups in *Sec. 3.4.1*; and the violent conflicts in the state of nature, including those among primitive humans in *Sec. 3.4.2*).

On the other hand, there is likewise technological transformation (e.g., the technical invention of vehicles for humans to move faster in *Sec. 2.2.1*; the technological construction of a metropolitan area for urban density in *Sec. 2.2.2*; the technological transformation of urban space to balance with nature for sustainable urban planning in *Sec. 2.3*; the technological construction of AFVs for humans to move around faster and more cheaply in *Sec. 2.4.3*; the technical invention of man-made sounds "by machines and equipment, such as radios, stereos,…television sets, motorboats, cars, motor or trail bikes, snowmobiles or chain saws," in *Sec .2.5.2*; the anthropogenic causes of deforestation, especially with the use of technology for mining, infrastructure construction, and farming, for instance, in *Sec. 3.3.1*; the use of technological means to construct urban open spaces in our times in *Sec. 3.3.2*; the use of technology, as in the building of railroads and automobiles, for the suburban way of life, in *Sec .3.4.1*; and the increasing use of technology in urban warfare, like precision-guided bombs or bulldozers, in *Sec. 3.4.2*).

Again, it should be stressed, however, that there are different shades of gray (or different degrees of truth) in the two opposites, and the classification is not necessarily mutually exclusive either. In fact, even when some combinations of the two occur, they only end up sharing the same dialectic relationship, but in a different degree.

Also, there are some other relationships (as a kind of family resemblance) which have something in common with the principle, but they are not exactly the same but only more or less comparable, which varies from case to case.

Good examples of family resemblance in relation to the principle are inorganicness-volition (or unwillingness-willingness), and whatnot.

## 5th Thesis: The Symmetry-Asymmetry Principle

The fifth thesis is labeled as the symmetry-asymmetry principle (on the relationships among existents) in the category of "agency" in existential dialectics (which was also already worked out in my previous books), in that there is no asymmetry without symmetry.

For instance, the Same can be symmetric and asymmetric towards the Others. But in case of asymmetry, oppression and self-oppression can occur. So, when the Same is asymmetric towards the Others, the Same can also be relatively asymmetric towards itself in self-oppression, just as the Others can be likewise towards themselves. The subsequent oppressiveness is dualistic, as much by the Same against the Others and itself, as by the Others against the Same and themselves. Both oppression and self-oppression can be achieved by way of downgrading differences between the Same and the Others and of accentuating them.

This is true, even though not all forms of asymmetry have to be about oppression and self-oppression.

In addition, from Chapter Three of *FPHG*, symmetry is not perfect, to be understood in an approximate sense under many life circumstances. With this clarification in mind, hereafter are some empirical examples.

For instance, on the one hand, there is symmetry (e.g., the co-existence of different groups living together in an urban space in *Sec .2.2.2*; the co-existence of different participants with their diverse opinions and emotions on LEK in *Sec. 2.3*; the co-existence of different approaches to fight urban pollution in *Sec. 2.4.1*; the co-existence of both the rich and the poor in urban areas in *Sec. 2.4.2*; the co-existence of different vehicles, both gasoline-fuel and alternative-fuel in *Sec. 2.4.3*; the co-existence of different religious values in urban planning in *Sec. 2.5.1*; the co-existence of different sounds in *Sec. 2.5.2*; the co-existence of both "face-to-face" relations and "side-by-side" relations in urban life in *Sec. 3.2.1*; the co-existence of different views on the causes of urban homelessness in *Sec. 3.2.2*; the co-existence of natural and anthropogenic deforestation in *Sec. 3.3.1*; the co-existence of different residents in a city to enjoy urban open spaces in *Sec. 3.3.2*; the co-existence of urban decay and urban renewal in *Sec. 3.4.1*; the co-existence of conventional warfare and urban warfare in *Sec .3.4.2*; the co-existence of different individualist and communitarian views of personal autonomy in the context of modern urban life in *Sec. 3.5.1*; and the co-existence of different sects or demoninations of Buddhism in *Sec. 3.5.2*).

On the other hand, there is also asymmetry (e.g., the possibility of urban flight for those with higher income levels, but the trap of urban slums for those with lower income levels, in *Sec .2.2.2*; the use of LEK from some of the participants and the rejection of LEK from others in *Sec. 2.3*; the popularity of the Coase's theorem among free marketers, and the acceptance of government intervention among those on the Left, in *Sec. 2.4.1*; nice neighborhoods for the rich and bad ones for the poor in *Sec. 2.4.2;* the popularity of gasoline-fuel vehicles in the older days, and the increasing preference for AFVs in our time in *Sec. 2.4.3*; the popularity of Christian religion in medieval urban planning, and the fashionable acceptance of rationalism in modern urban planning in *Sec. 2.5.1*; the pleasantness of some sounds for certain individuals, but the unpleasantness of other sounds for them, in *Sec. 2.5.2*; "some urban subjects are more invisible, more anonymous than others," in that "side-by-side" relations can be had for some, while the others opt for the "face-to-face" relations, in *Sec. 3.2.1*; the political correctness in our time to not emphasize too much on the bio-psychological causes of urban homelessness, as opposed to the older days when the homeless could not more easily regarded as suffering from madness, in *Sec. 3.2.2*; the more serious impact of natural deforestation in antiquity, but the more serious scale of anthropogenic deforestation in our time, in *Sec. 3.3.1*; the benefits of urban open spaces more for the well-to-do but less for the homeless in *Sec. 3.3.2*; the worse impact of urban decay for the poor, and the better impact of urban renewal for the rich, in *Sec. 3.4.1*; the popularity of conventional warfare among nation-states, but the relevance of urbicide in urban warfare, in *Sec. 3.4.2*; the acceptance of individualism by individuals like Emerson, and the popularity of communitarianism by individuals like Sandel, in *Sec. 3.5.1*; and the popularity of Zen Buddhism in Japan, but the acceptance of Theravada Buddhism in Southeast Asia, in *Sec. 3.5.2*).

Once more, it should be stressed, however, that there are different shades of gray (or different degrees of truth) in the two opposites, and the classification is not necessarily mutually exclusive either. In fact, even when some combinations of the two occur, they only end up sharing the same dialectic relationship, but in a different degree.

Also, there are some other relationships (as a kind of family resemblance) which have something in common with the principle, but they are not exactly the same but only more or less comparable, which varies from case to case.

Good examples of family resemblance in relation to the principle are strength-weakness, and something like that.

## 6th Thesis: The Regression-Progression Principle

The sixth thesis is referred to as the regression-progression princple (on the direction of history) in the category of "outcome" in existential dialectics (which was also already worked out in my previous books), in that neither the cyclical nor the linear views are adequate for explaining many phenomena at all levels.

History progresses to more advanced forms, but with a regressive touch. Examples include no freedom without unfreedom, no equality without inequality, and no civilization without barbarity. This is not an inevitable law, but merely a highly likely empirical trend.

For instance, on the one hand, there is progress (e.g., the construction of roads for faster movement from one place to another in *Sec. 2.2.1*; the development of metropolitan areas to accommodate the rise of urban population over time in *Sec. 2.2.2*; the better understanding of ecological interdependence over time in *Sec. 2.3*; the better understanding of the seriousness of urban pollution and the available means to counter it in *Sec. 2.4.1*; the awareness of the need for urban renewal with the consideration of the plight of minorities and the ecological environment in *Sec. 2.4.2*; the invention of AFVs in *Sec. 2.4.3*; the design of modern buildings to confront the challenge of ever increasing urban density, as in the works of Wright and Corbusier, in *Sec. 2.5.1*; the implementation of soundscape planning to fight noise pollution in *Sec. 2.5.2*; the building of different forms of community in urban density in *Sec. 3.2.1*; the numerous efforts by both federal and local governments to provide different aids to the homeless, like shelters and psychiatric services, in *Sec. 3.2.2*; the increase of food production worldwide by way of deforestation in *Sec. 3.3.1*; the construction of urban open spaces as an escape from life stress in urban density in *Sec. 3.3.2*; the efforts for urban renewal by way of gentrification and planned shrinkage in *Sec. 3.4.1*; the better way to fight urban warfare by urbicide than by the conventional Clausewitzian means in *Sec. 3.4.2*; the individualist rejection of the violent history of "religious wars, ethnic hatred, punishing discrimination, callous oppression, and the like" for a better lifeworld, in *Sec. 3.5.1*; and the Buddhist contribution to the freedom from urban suffering by way of the monastic life in urban void in *Sec. 3.5.2*).

On the other hand, there is regression (e.g., the occurrence of traffic jam and road rage in *Sec. 2.2.1*; the dark side of urbanization, such as urban decay and social breakdown in *Sec. 2.2.2*; the lack of general agreement on collecting and using LEK, with the subsequent serious questioning on the extent of possibility and desirability of sustainability in urban planning, in *Sec. 2.3*; the collusion among politicians, regulators, and interest groups, as shown in public choice theory, in *Sec. 2.4.1*; the persistent gap between the rich and the poor in urban habitation, and the serious recurrence of ecological problem, in *Sec. 2.4.2*; the failure of many AFV markets in the last century, and its unintended policy impacts, in *Sec. 2.4.3*; the religious bias against sexual deviance and disorder in modern urban planning in *Sec. 2.5.1*; the costly fight among different groups for any soundscape planning, often without uneven appreciation of the marginal views of the disadvantaged, like the minorities, the poor, or the hearing-impaired, in *Sec. 2.5.2*; the emergence of urban solitude for those who do not fit in, with the subsequent consequences of "aversion, disregard or neglect," in *Sec. 3.2.1*; the persistence of urban homelessness over time, due to bio-psychological causes like mental illness and others, in *Sec. 3.2.2*; the costly impacts of deforestation, be they atmospheric, hydrologic, soil, economic, and ecological, in *Sec. 3.3.1*; the subsequent problems associated with the construction of urban open spaces, such as urban sprawls, segmentation, class bias, etc., in *Sec. 3.3.2*; the actual worsening of urban decay in some other aspects of gentrification and planned shrinkage, in *Sec. 3.4.1*; the brutal outcome of urbicide in urban warfare in *Sec. 3.4.2*; the feverish obsession, solipsism, and loneliness in the individualist self, in *Sec. 3.5.1*; and the dark sides of Buddhism in practice, be they about boredom, exclusiveness, and indulgence, in *Sec. 3.5.2*).

Again, it should be stressed, however, that there are different shades of gray (or different degrees of truth) in the two opposites, and the classification is not necessarily mutually exclusive either. In fact, even when some combinations of the two occur, they only end up sharing the same dialectic relationship, but in a different degree.

Also, there are some other relationships (as a kind of family resemblance) which have something in common with the principle, but they are not exactly the same but only more or less comparable, which varies from case to case.

Good examples of family resemblance in relation to the principle are decline-growth, and so forth.

## 7th Thesis: The Post-Human Planning

The seventh thesis concerns the role of "post-humans," which I originally proposed in my previous books, starting with the first book titled *The Future of Human Civilization* in 2000 and all others afterwards.

As already pointed out in *Sec. 1.7*, I need to emphasize, as this is something that I used to repeat (*almost verbatim*) from my previous books, two clarifications here about the term "post-human" as a neologism in my works.

Firstly, the word "post-human" here should *not* be confused with another term which looks similar but has a totally different meaning in the literature of postmodernism, namely, "post-humanism"—which constitutes a critique of "humanism" as tradionally understood (especially, though not exclusively, in relation to the idea of progress in science and reason in the Enlightenment project). (WK 2008)

My works reject the project of "postmodernism" and propose the future world of what I orignally called "after-postmodernity" in *FHC* and *FCD*, for instance.

And secondly, the word "post-human" here should also *not* be confused with a similar term which is used to champion the ideology of technology for the future co-existence between humans and cyborgs in "trans-humanism." (WK 2008a)

Instead, my term "post-human" in relation to "posthuman-ism" also rejects "transhumanism" (especially, though not exclusively, in relation to the promises of technology) and refers to something else altogether, that is, the future extinction of humans and its post-human successors in deep space and beyond unto multiverses.

With these two clarifications in mind (as summarized in *Table 1.26*)—the post-human urban planning will likely take two main paths, based on the analysis in this book so far (and also on the visions as already worked out in my previous books).

Firstly, the "post-human" urban planning will make good use of the contingent theory of urban planning in two comprehensive ways, namely, (a) in relation to the two realms of density and void, and (b) in relation to the four perspectives of the mind, nature, society, and culture.

And secondly, the "post-human" urban planning will likely evolve, in relation to my visions on the mind (e.g., "floating consciousness," "hyper-spatial consciousness," "unfolding unconsciousness," "thinking machines," "thinking robots," "genetically altered superior beings," "cyborgs," and others), nature (e.g., the "alteration of space-time," the "creation of new matter-energy," and the like), society (e.g., different versions of "post-capitalism" and "post-democracy," the movement of "cyclical progression," "ambivalent technology," and so forth), and culture (e.g., the rise of "post-civilization," "transformative aesthetic experience," "contrastive mathematical-logic," and the whatnot) in history (e.g., the age of "after-postmodernity"), in the context of my approaches in relation to methodology (e.g., "sophisticated methodologial holism") and ontology (e.g., "existential dialectics").

Many, though not all, of my visions on the mind, nature, society, culture, history, methodology, and ontology are summarized in the tables as shown in Chapter One and Chapter Four.

With these visions in mind—what then is the implication for the future of sustainability?

## Beyond Sustainsability

In the end, the post-humans will likely consider sustainbility as an intellectual fad in our times, in light of the critical analysis of the very extent of the possibility and desirability of sustainability in the previous chapters, even if the focus on urban planning is used as a case study here.

Contrary to contemporary wisdom, the realization of sustainability in urban planning, in a small sense—and in other serious challenges in our times (e.g., ecological, economic, demographic, technological, moral, spiritual, and the like), in a broad sense—presupposes some conditions about the human mind, nature, culture, and society, which, when closely examined, are neither possible nor desirable to the extent that its proponents would so often like us to believe.

After all, nothing lasts foreever anyway, neither can sustainability. Each urban design for a new city appears and becomes popular for a while in history, just as it then fades away as its historical usefulness expires, to be replaced by a new one in a different era later. Thus continues the same cycle of change.

This is not to say that one should therefore dismiss the issue of sustainability once and for all. Of course not, the business of opting for optimality (or at least, something close to it) in dealing with complex problems in the realm of sustainability will continue, but the point here is that it will not be possible and desirable to the extent that its proponents are often so ready to claim.

Surely, in the long-haul, the post-humans will further transform the very enterprise of urban planning (or, sustainability, in a broad sense) in their own distinctive directions more suitable to the needs and interests of the future eras. In the process, they will no doubt strive for a different version of optimality (or something close to it) beyond our current technological imagination. Even then, the outcome will likely be subject to the constraints as revealed in the ontological principles of existential dialectics.

With this note of caution in mind, it is time to go beyond the current fashionable discourse on sustainability and start thinking about a future world, of which this book has only provided a rough historical sketch, in conjunction with the helpful summary of all related previous books of mine.

And the rest will be taken care of by posterity—and, for that matter, the post-human successor in future history.

**Table 4.1. Sophisticated Methodological Holism (Part I)**

• "My methodological holism implies the partiality-totality principle in the ontology of existential dialectics (see the table on the partiality-totality principle for summary), which is against the varieties of (a) reductionism and (b) reverse-reductionism, in relation to (i) concept, (ii) theory, (iii) methodology, and (iv) ontology." (*FC*)

• "[M]y methodological holism here is not opposed to methodological individualism but includes it (and, for that matter, other methodologies too)…." (*FPHC*) For this reason (and others too, as summarized hereafter), my version of methodological holism is sophisticated—not vulgar as sometimes used by inapt scholars using the same term. (*FC*)

• "[M]y methodological holism does not democratically presume that all levels are equally valid, as all levels are not created equal. In other words, in relation to issue X, level A may be more relevant than level B, but in relation to Y, level B can be more relevant than level A instead." (*FPHC*) One excellent example of this vulgar democratic presumption is what I called in *BNN* "the compromise fallacy." (*FC*)

• My methodological holism does not presume that a lower level of analysis is more important than a higher level, solely because the former serves as the foundation for the latter—and vice versa, for that matter. One excellent example of this reductionistic presumption is what I called in *FPHST* "the foundation fallacy." (*FPHST*, *FC*)

*(continued on next page)*

**Table 4.1. Sophisticated Methodological Holism (Part II)**

• "[M]y methodological holism does not make any a-priori postulation that there must be a definite (and, for that matter, indefinite) number of levels" in any analysis. (*FPHC*) Nor does it dogmatically require that there must be a certain combination of levels of analysis in a given inquiry. (*FC*)

• "[M]ethodological holism, in my usage, does not assume that all levels…can necessarily be integrated, since methodological holism is not aimed to search for the holy grail of 'an integral theory…' (as is the case for Wilber). In other words, it allows that sometimes some levels may experience irreducible gaps between them, to be understood, at best, as empirical correlations, not as causal relations…." (*FPHC*)

• "[D]ifferent levels may overlap and even interact with each other in a given context (but sometimes may not), and the fact that I even proposed different ways of re-classifying the levels (whenever needed) in *FDC* reinforces this point….The dual danger here is either forcefully making different levels interact when they are just different (or, metaphorically speaking, apples and oranges) or inappropriately ignoring their interactions when some situations instead require them." (*FPHC*)

• "[T]o understand different levels from their own (unique) perspectives (as required by my methodological holism) is not the same as trying to reduce them to a preferred level in the process of learning from other levels. This second kind of multidisciplinary work is not genuine and does no justice to the unique complexities and merits inherent at each level." (*FPHC*)

*(continued on next page)*

**Table 4.1. Sophisticated Methodological Holism (Part III)**

• "[My] methodological holism walks a fine line between the artificial classification (separation) of levels and the simultaneous incorporation of them, if only for the sake of human scholarly endeavor. It should be reminded that nature does not impose upon itself the academic classification of the levels of analysis as humans have. The enterprise of classification is therefore anthropocentric." (*FPHC*)

• "[M]y methodological holism advocates neither epistemic subjectivism nor epistemic non-subjectivism (e.g., realism, idealism, and historicism), neither epistemic relativism (e.g., subjectivism, historicism) nor epistemic absolutism (e.g., realism, positivism), neither epistemic reductionism nor epistemic emergencism, and neither epistemic objectivism (e.g., realism, idealism) nor epistemic historicism….Neither does methodological holism, in my usage, accept the false meta-conceptual dichotomy between nominalism and realism….These false dichotomies…are to be transcended. In other words, methodological holism does not fully accept epistemic realism, positivism (a form of epistemic idealism), historicism, subjectivism, and reductionism in epistemology and philosophy of science but learns from the strengths and weaknesses of all of the opposing approaches without siding with any of them…." (*FPHC*)

• "Sophisticated methodological holism is subject to the constraints as imposed by the syntax of existential dialectics (e.g., the partiality-totality principle and the predictability-unpredictability principle). Even in predictability, outcomes are subject to uncertainty, the degree of which varies from case to case." (*FC*)

*(continued on next page)*

**Table 4.1. Sophisticated Methodological Holism (Part IV)**

• "Sophisticated methodological holism—when applied, especially though not exclusively, as illustrated in my numerous works—can enrich the understanding of reality in some distinctive ways. Here are three examples (as revealed in each of my books)." (*FPHCT*)

—"Firstly, it provides a comprehensive analysis of a subject matter, from which much can be learned about reality, in relation to the perspectives of the mind, nature, society, and culture."

—"Secondly, it suggests some visions of the future in relation to the subject matter in question."

—"And thirdly, it proposes some insights on meta-theory (e.g., methodology and ontology) in general—with the clear understanding, however, of the dilemma of specific vs. general ontology (as shown in the table on the syntax of existential dialectics in the context of the dilemma of ontology). For this very reason, all of these ways are important, without reducing one into the analysis of another."

*Sources*: A summary of *Sec.1.2* in *FPHC*—and also from *BNN*, *FPHST*, *ALD*, *FC*, *FPHCT*, and others. See the books for more detail.

**Table 4.2. On Reductionism and Reverse-Reductionism (Part I)**

---

• **The Partiality-Totality Principle**

—The partiality-totality principle in the ontology of existential dialectics targets against the varieties of reductionism and reverse-reductionism (as already worked out in my previous books).

• **Against the Varieties of Reductionism**

—*Conceptual Reductionism*

- Some illustrative instances involve myriad dualities like mind vs. body, self vs. world, democracy vs. non-democracy, and the like (as already addressed in *FHC*, *FPHC*, and *BDPD*, for instance).

—*Methodological Reductionism*

- A good instance concerns the debate between different versions of qualitative and quantitative methods (as already analyzed in *FC* and also *FHC*). In *FPHML*, I examined similar reductionism, this time, in the literature on mathematical logic (e.g., the obsession with consistency, soundness, and completeness).

—*Ontological Reductionism*

- An excellent example is the debate between emergentism and reductionism in complexity theory and also in psychology (as elaborated in *FPHC*, in the context of Being and Becoming).

---

*(continued on next page)*

**Table 4.2. On Reductionism and Reverse-Reductionism (Part II)**

- **Against the Varieties of Reductionism** ***(cont'd)***
  —*Theoretical Reductionism*
    - A fascinating case study concerns what I originally called "the foundation fallacy" in *FPHST*, in any attempt to naively understand space-time from the physical perspective as the foundation and, consequently, to dangerously dismiss other perspectives.
    - In *FAE*, I elaborated further these versions of reductionism in the literature on aesthetics (e.g., form vs. content, representation vs. expression, critics vs. artists, and externalism vs. internalism).
    - In *FIA*, I revealed other forms of reductionism in the literature on information architecture (e.g., the constructivist argument).
    - In *FPHU*, I showed the persistent legacy of reductionism, this time, in the literature on anomalous experience (e.g., the obsession with physics, chemistry, and biology for explaining anomalous experience).
    - In *FPHE*, I examined another case of reductionism in action, in the context of engineering (e.g., technical constraints vs. normative constraints).
    - In *FPHMM*, I elaborated one more version of reductionism, in relation to the three domains of communication (e.g., the competing views on sending, connecting, and receiving).
    - In *FPHCT*, I explored another version of reductionism, in relation to invention and innovation (e.g., the bio-psychological argument vs. the socio-cultural arguments).
    - In *FPHG*, I identify another version of reductionism, in relation to infinity, symmetry, and dimensionality (e.g., the Euclidean argument vs. the non-Euclidean arguments).
    - And in *FPHUP*, I analyzed another version of reductionism, in relation to density and void (e.g., the engineering argument and the ecology argument).

*(continued on next page)*

**Table 4.2. On Reductionism and Reverse-Reductionism (Part III)**

---

- **Against the Varieties of Reverse-Reductionism**

—*Conceptual* Reverse-Reductionism

- Any concept of "art" (e.g., fine art, cave art, outsider art, junk art) is deemed acceptable in postmodernism (as already addressed in Ch.4 of *FHC*).

—*Theoretical Reverse-Reductionism*

- There are numerous art and literary theories co-exist. Take the case of literature, as there are now Literary Structuralism, Marxist Literary Criticism, New Criticism, Phenomenology, Hermeneutics, Language-Game Literary Criticism, Feminist Literary Criticism, Reception Theory, Reader Response Criticism, Poststructuralism, Semiotics, Pyschoanalytic Literary Criticism, just to cite some well-known ones, with no one being said to be better than any others (as detailedly analyzed in Ch.4 of *FHC*). (S. Raman 1997) In *BNN*, I even introduced "the compromise fallacy" as another good illustration of theoretical reverse-reductionism, in misleadingly treating both genetic and environmental approaches as equally valid.

—*Methodological Reverse-Reductionism*

- There is the "anything-goes" mentality in postmodernism (e.g., doing art without praxis, doing art with praxis, and doing art by sublation), as analyzed in Ch.4 of *FHC*.

—*Ontological Reverse-Reductionism*

- There are likewise no privileged ontology, and the door is open for anything in postmodernism (e.g., the equal status of the ontology of Being vs. that of Becoming, as already addressed in Ch.4 of *FHC*—and also in *FPHC).* In *FAE*, I also introduced another version of reverse-reductionism, that is, "the pluralist fallacy," in the context of understanding aesthetic experience, for instance—although this fallacy has been committed not exclusively in relation to the ontological level (but also at the conceptual, theoretical, and methodological ones).

---

*(continued on next page)*

**Table 4.2. On Reductionism and Reverse-Reductionism (Part IV)**

*Sources*: A summary of my previous books, especially though not exclusively *FHC*, *FCD*, *FPHC*, *BDPD*, *BCPC*, *BNN*, *FPHST*, *BCIV*, *FC*, *FAE*, *ALD*, *FIA*, *FPHU*, *FPHML*, *FPHE*, *FPHMM*, *FPHCT*, *FPHG*, and *FPHUP*

**Table 4.3. The Concepton of Existential Dialectics (Part I)**

- **Sets and Elements**
  - —Sets
    - Ex: the Same
    - Ex: the Others
  - —Elements
    - Ex: whites in 20th century America (in the set of "the Same")
    - Ex: Iraq during the U.S. invasion in 2003 (in the set of "the Others")

- **Relations, Operations, Functions**
  - —Relations (e.g., "belongs," "equals to," "is greater than")
    - Ex: symmetric interactions within the Same (or the Others)
    - Ex: asymmetric interactions between the Same and the Others
  - —Operations (e.g., "and," "or," "not," "if…then")
    - Ex: if the Same oppresses the Others, it will also oppress itself.
    - Ex: the Same is not the Others.
  - —Functions (e.g., goals)
    - Ex: the Same is hegemonic in relation to the Others.

- **Truth Values**
  - —"1" if True (in Symbolic Logic)
    - Ex: the proposition that imperial Japan was hegemonic to China during WWII
  - —"0" if False (in Symbolic Logic)
    - Ex: the proposition that Grenada invaded France in 2003
  - —"1" & "0" if Both True and False (in Dialectic Logic)
    - Ex: the proposition that the rabbit-duck picture refers to a duck
  - —"~1" & "~0" if Neither True Nor False (or N/A, in Dialectic Logic)
    - Ex: the proposition that God really exists

*(continued on next page)*

**Table 4.3. The Conception of Existential Dialectics (Part II)**

- **Axioms, Postulates, Theorems, Principles**
  - —Axioms
    - Ex: the reflexive axiom—"any quantity is equal to itself"
  - —Postulates
    - Ex: the SSS postulate—"if the three sides of a triangle are congruent to their corresponding parts, then the triangles are congruent"
  - —Theorems (and Principles) in Existential Dialectics
    - <u>In Relation to Method</u>
      - –#1: the formalness-informalness principle
      - –#2: the absoluteness-relativeness principle
      - –#3: the partiality-totality principle
      - –#4: the predictability-unpredictability principle
      - –#5: the explicability-inexplicability principle
    - <u>In Relation to Structure</u>
      - –#6: the finiteness-transfiniteness principle
      - –#7: the preciseness-vagueness principle
      - –#8: the simpleness-complicatedness principle
      - –#9: the openness-hiddenness principle
      - –#10: the denseness-emptiness principle
    - <u>In Relation to Process</u>
      - –#11: the change-constancy principle
      - –#12: the order-chaos principle
      - –#13: the slowness-quickness principle
      - –#14: the expansion-contraction principle
    - <u>In Relation to Agency</u>
      - –#15: the theory-praxis principle
      - –#16: the convention-novelty principle
      - –#17: the evolution-transformation principle
      - –#18: the symmetry-asymmetry principle
      - –#19: the softness-hardness principle
    - <u>In Relation to Outcome</u>
      - –#20: the regression-progression principle
      - –#21: the same-difference principle

*(continued on next page)*

**Table 4.3. The Conception of Existential Dialectics (Part III)**

*Notes*: The categories and examples in each are solely illustrative (not exhaustive). The comparison is also relative (not absolute), nor are they mutually exclusive. As generalities, they allow exceptions.
*Sources*: Starting from Ch.6 of *BCPC* and also from other books of mine

**Table 4.4. The Syntax of Existential Dialectics I: The Principles (Part I)**

- **In Relation to Method**

—*The Formalness-Informalness Principle*
*(On the Formal Requirements of Logical Systems)*

- The formal requirements of a logical system (e.g., consistency, soundness, and completeness) have both usefulness and non-usefulness, to the extent that, if there are formal systems requiring them, there are alternative ones which do not. Thus, it does *not* exclude classical logics but simply goes beyond both classical and non-classical logics, while learning something from each.
- For instance, existential dialectics can make use of both classical logics under certain conditions (e.g., especially, though not exclusively, when they are clear-cut, etc.) and non-classical logics under alternative conditions (especially, though not exclusively, when they are "unknown," "irrelevant," "ambiguous," "possible," with "different degrees of truth," empirically inconsistent in a desirable way, etc.). (WK 2008u)
- Family resemblance: e.g., logicalness-nonlogicalness
- Sources: From *FPHML*. See also other books of mine.

—*The Explicability-Inexplicability Principle*
*(On the Underlying Mechanisms of Things)*

- Both explicability and inexplicability are part of the understanding of things.
- This principle tells us the duality of the research dilemma, in that, if reality can be explained in some ways, it also has its other ways which are not quite explainable, at a given point in time.
- Family resemblance: e.g., underlyingness-regularness
- Sources: Especially from *FPHU*—and also *FPHC*. See also *FHC*, *FCD*, *BCPC*, *FPHST*, *BCIV*, *BNN*, *BWT*, *FAE*, *ALD*, *FPHK*, *FPHML*, *FPHE*, *FPHMM*, *FPHCT*, *FPHG*, & *FPHUP*

*(continued on next page)*

**Table 4.4. The Syntax of Existential Dialectics I: The Principles (Part II)**

• **In Relation to Method *(cont'd)***

—*The Predictability-Unpredictability Principle*
*(On the Occurrence of Events)*

- Both predictability and unpredictability have a major role to play in the occurrence of things, so that neither determinism nor indeterminism wins the centuries-old fight.
- There are events which are predictable, just as there are those which are not. Or what is regarded as unpredictable at one point in time may turn out to be predictable later, and, conversely, what is deemed as predictable may turn out to not be so predictable. Even in predictability, outcomes are subject to uncertainty, the degree of which varies from case to case.
- Family resemblance: e.g., sureness-arbitrariness
- Sources: Especially from *FC*. See also *FHC*, *FCD*, *FPHC*, *BCPC*, *FPHST*, *BCIV*, *BNN*, *BWT*, *FAE*, *ALD*, *FIA*, *FPHU*, *FPHK*, *FPHML*, *FPHE*, *FPHMM*, *FPHCT*, *FPHG*, & *FPHUP*

—*The Partiality-Totality Principle*
*(On the Relationships between Whole and Parts)*

- The whole is not the sum of the parts.
- Any inquiry about a phenomenon in the work is to guard against the varieties of (a) reductionism and (b) reverse-reductionism.
- Reductionism and reverse-reductionism can be (i) conceptual, (ii) theoretical, (iii) methodological, and (iv) ontological.
- Family resemblance: e.g., individualisticness-holisticness
- Sources: Especially from *FC*. See also *FHC*, *FCD*, *FPHC*, *BCPC*, *FPHST*, *BCIV*, *BNN*, *BWT*, *FAE*, *ALD*, *FIA*, *FPHU*, *FPHK*, *FPHML*, *FPHE*, *FPHMM*, *FPHCT*, *FPHG*, & *FPHUP*..

*(continued on next page)*

**Table 4.4. The Syntax of Existential Dialectics I:**
**The Principles**
**(Part III)**

---

• **In Relation to Method *(cont'd)***

—*The Absoluteness-Relativeness Principle*
*(On the Multiplicity of Things)*

- There is the multiplicity of things in reality, be they about entities, qualities (or properties), and relationships, such that what is acceptable from one standpoint may not be so from another. For instance, if there is something absolute, there is likewise something relative.
- Both absoluteness and relativeness here are also relevant to different modalities often cited in the literature on ontology, such as possibility (e.g., something "can" happen) and its opposite (e.g., impossibility), probability (e.g., something "will" happen) and its opposite (e.g., improbability), and necessity (e.g., something "should" happen) and its opposite (e.g., contingency).
- Family resemblance: e.g., uniformity-diversity
- Sources: From *FPHK*. See also *FHC*, *FCD*, *FPHC*, *BCPC*, *FPHST*, *BCIV*, *BNN*, *BWT*, *FAE*, *FIA*, *ALD*, *FPHU*, *FPHML*, *FPHE*, *FPHMM*, *FPHCT*, *FPHG*, & *FPHUP*.

---

*(continued on next page)*

**Table 4.4. The Syntax of Existential Dialectics I: The Principles (Part IV)**

- **In Relation to Structure**

—*The Finiteness-Transfiniteness Principle*
*(On the Nature of Numbers)*

- If there are finite things, there are likewise transfinite ones.
- To avoid confusion, my usage of the word "transfinite" here differs radically from the one used by Cantor (and other mathematicians) for "relative" infinity—and is more limited, in light of the problems confronting any attempt to understand the idea of infinity, be it by intuition, imagination, and conception (as detailedly analyzed in *Sec. 2.2.3* of *FPHG*).
- Instead, by "transfinity," I allow numbers which can be many times larger—or smaller, for that matter—than the finite things that we encounter in daily life, but they do not have to be related to the idea of infinity at all (which may not exist).
- Of course, there may be some *borderline* cases, in which it is not clear whether the number in question is transfinite (in my usage) or simply a mathematical convenience. A good example of a borderline case is the Planck unit of length for "the smallest space possibly measured in nature," which is "less than billionths of trillionths of trillionths of an inch" (or something like $1.6 \times 10^{-35}$ meters).
- That said—my usage of "transfinity" can also resolve (or better, dissolve) an age-old problem in philosophy known as "Zeno's paradoxes" (as already explained in *Sec. 2.2.3* of *FPHG*).
- Family resemblance: e.g., boundedness-quasiunboundedness
- Source: From *FPHG*. See also *FPHUP*.

*(continued on next page)*

**Table 4.4. The Syntax of Existential Dialectics I:**
**The Principles**
**(Part V)**

- **In Relation to Structure** ***(cont'd)***

—*The Simpleness-Complicatedness Principle*
*(On the Interconnection among Things)*

- Both simpleness and complicatedness are vital, without favoring one over the other, in that each is utilized, depending on the basis of the perspectives of nature, the mind, culture, and society. And even when a combination of them is preferred, the dilemma is only shifted to a combinational degree of concern.
- In relation to taxonomy, simpleness has its heuristic usefulness, just as complicatedness has its realistic representation, for instance. And in relation to network, simpleness has its economical attractiveness, just as complicatedness has its practical reliability, for instance.
- Family resemblance: e.g., inflexibility-flexibility, standardization-specialization
- Sources: Especially from *FIA*. See also *FPHU*, *FPHK*, *FPHML*, *FPHE*, *FPHMM*, *FPHCT*, *FPHG*, & *FPHUP*.

—*The Preciseness-Vagueness Principle*
*(On the Refinement of Things)*

- Both preciseness and vagueness are important, not that one is better than the other, but that both are used, in different degrees of preference, in accordance to the contextual application from the perspectives of nature, the mind, culture, and society. Even when both are used in a combination, the dilemma is shifted instead to one of combinational concern.
- In relation to taxonomy, preciseness has its taxonomic clarity, just as vagueness has its classificatory flexibility, for instance. And in relation to network, vagueness has its explorative liberty, just as preciseness has its conceptual definitiveness, for instance.
- Family resemblance: e.g., clarity-ambiguity, directness-indirectness
- Sources: Especially from *FIA*. See also *FPHU*, *FPHK*, *FPHML*, *FPHE*, *FPHMM*, *FPHCT*, *FPHG*, & *FPHUP*.

*(continued on next page)*

**Table 4.4. The Syntax of Existential Dialectics I: The Principles (Part VI)**

- **In Relation to Structure** ***(cont'd)***

—*The Openness-Hiddenness Principle*
*(On the Detection of Things)*

- This principle suggests that reality has its hidden face, just as it is open to outside view in some other ways.
- For instance, in the context of anomalous experience, certain aspects of reality concerning anomalous experience can be open for examination (e.g., the experiment with SPECT images, the transmission of telepathemic bit, and the use of metonymies and metaphors). Other aspects, however, remain hidden, and examples include the elusive deeper nature of spiritual reality in the intangible realm, the undetectability of different branched universes, the underlying mechanisms of signals traveling between different folds of physical space-time, and the dependence on language and the bias for science and logic).
- Family resemblance: e.g., overtness-covertness
- Sources: Especially from *FPHU*. See also *FPHK*, *FPHML*, *FPHE*, *FPHMM*, *FPHCT*, *FPHG*, & *FPHUP*.

—*The Denseness-Emptiness Principle*
*(On the Distribution of Entities in Space)*

- Both density and void are needed, in relation to the mind, nature, culture, and society, albeit in different ways.
- Lest any misunderstanding occurs, the term "void" is used here only as an approximation of emptiness (depending on the degree of the lack of density), since, in physics, it is well known that "empty" space is not really empty all the way, because it can be full of energy (e.g., random quantum fluctuations at the sub-atomic level, and, for that matter, dark energy in the universe) and matter (e.g., different versions of sub-atomic particles, and, for that matter, dark matter in the universe).
- Family resemblance: e.g., fullness-void
- Sources: First worked out in *FPHUP*

*(continued on next page)*

**Table 4.4. The Syntax of Existential Dialectics I: The Principles (Part VII)**

• **In Relation to Process**

—*The Expansion-Contraction Principle*
*(On the Growth of Things)*

- Entities in the world can both expand in some ways and contract in other ones, as part of their nature
- For instance, in the context of anomalous experience, one of its most fascinating expressions is none other than the analysis of different causes of the slow but steady expansion (and, furthermore, if put in a different way, unfolding) of unconsciousness.
- Family resemblance: e.g., conquest-autarky
- Sources: Especially from *FPHU*. See also *FCD*, *FPHC*, *BWT*, *FPHK*, *FPHML*, *FPHE*, *FPHMM*, *FPHCT*, *FPHG*, & *FPHUP*.

—*The Slowness-Quickness Principle*
*(On the Speed of Change)*

- Both slowness and quickness co-exist, with their own internal tension, to the extent that each fights for its own relevance with the other, in accordance to the perspectives of nature, the mind, culture, and society, without one being the victor and the other being the vanquished in the long haul. Even when both are chosen in other cases, this dilemma is only transferred into something else with a combinational character.
- In relation to taxonomy, quickness has its efficient usability, just as slowness has its aesthetic appeal, for instance. And in relation to network, quickness has its adventurous readiness, just as slowness has its risk-adverse convenience, for instance.
- Family resemblance: e.g., inconvenience-convenience
- Sources: Especially from *FIA*. See also *FPHU*, *FPHK*, *FPHML*, *FPHE*, *FPHMM*, *FPHCT*, *FPHG*, & *FPHUP*.

*(continued on next page)*

**Table 4.4. The Syntax of Existential Dialectics I: The Principles (Part VIII)**

- **In Relation to Process** ***(cont'd)***

—*The Change-Constancy Principle*
*(On the Alteration of Things)*

- Change occurs over time, although constancy is also allowed.
- Asymmetry undergoes changes over time, so does symmetry.
- Old players fade away, and new ones emerges, with ever new causes and ever new forms.
- Family resemblance: e.g., dynamicness-staticness, instability-stability
- Sources: First named in *BCPC*. Especially from *FHC*, *FCD*, and *FPHC*. See also *BDPD*, *FPHST*, *BCIV*, *BNN*, *BWT*, *FC*, *FAE*, *ALD*, *FPHU*, *FPHK*, *FPHML*, *FPHE*, *FPHMM*, *FPHCT*, *FPHG*, & *FPHUP*.

—*The Order-Chaos Principle*
*(On the Pattern of Things)*

- Both order and chaos are vital in the process of change in the world. The preference for order is biased, since it does not give sufficient attention to the vital role of chaos in the transformation of the world (without somehow reducing it for the understanding of order).
- The scientific search for order in the world is often a hidden bias in its ontological obsession with order, since chaos is often treated as the "bad" guy, with order as the "good" guy (for the end goal of science).
- Neither order nor chaos is the final end of the world, and one is not to be treated as the means for the other in the transformation of things. Both are fundamental in their recurrent dialectical interactions with each other over time, without reducing one for the other.
- Family resemblance: e.g., lawfulness-disorder
- Sources: Especially from *FC*. See also *FHC*, *FCD*, *FPHC*, *BCPC*, *FPHST*, *BCIV*, *BNN*, *BWT*, *FAE*, *ALD*, *FIA*, *FPHU*, *FPHK*, *FPHML*, *FPHE*, *FPHMM*, *FPHCT*, *FPHG*, & *FPHUP*.

*(continued on next page)*

**Table 4.4. The Syntax of Existential Dialectics I:
The Principles
(Part IX)**

---

• **In Relation to Agency**

—*The Softness-Hardness Principle*
*(On the Force of Change)*

- This has to do with *the force of change* under the category about agency in existential dialectics, in that any change by an agent, be it organic (like humans) or non-organic (like natural objects), can occur in a forceful (aggressive) or gentle (pacific) way, which can come in all shapes and sizes, of course.
- Family resemblance: e.g., peacefulness-violence, cooperation-competition
- Sources: Especially from *ALD*. See also *FHC*, *FCD*, *FPHC*, *BCPC*, *FPHST*, *BCIV*, *BNN*, *BWT*, *FAE*, *FIA*, *FPHU*, *FPHK*, *FPHML*, *FPHE*, *FPHMM*, *FPHCT*, & *FPHG*.

—*The Theory-Praxis Principle*
*(On the Duality of Knowledge))*

- This has to do with the duality of knowledge in existential dialectics, in that, if there is theoretical construction, there is likewise its practical application, both technical and normative.
- Family resemblance: e.g., discovery-application
- Source: Especially from *FPHE*. See also *FPHMM*, *FPHCT*, *FPHG*, & *FPHUP*.

---

*(continued on next page)*

**Table 4.4. The Syntax of Existential Dialectics I:**
**The Principles**
**(Part X)**

---

• **In Relation to Agency *(cont'd)***

—*The Symmetry-Asymmetry Principle*
*(On the Relationships among Existents)*

• There is no symmetry without asymmetry.
• For instance, the Same can be symmetric and asymmetric towards the Others. But in case of asymmetry, oppression and self-oppression can occur. So, when the Same is asymmetric towards the Others, the Same can also be relatively asymmetric towards itself in self-oppression, just as the Others can be likewise towards themselves.
• The subsequent oppressiveness is dualistic, as much by the Same against the Others and itself, as by the Others against the Same and themselves.
• Both oppression and self-oppression can be achieved by way of downgrading differences between the Same and the Others and of accentuating them.
• This is true, even though not all forms of asymmetry have to be about oppression and self-oppression.
• In addition, from Chapter Three of *FPHG*, symmetry is not perfect, to be understood in an approximate sense under many life circumstances.
• Family resemblance: e.g., strength-weakness
• Sources: From all my books, starting with *FHC*. First named in *BCPC*. See, for instance, *FCD* for more details on the Same and the Others.

---

(*continued on next page*)

**Table 4.4. The Syntax of Existential Dialectics I: The Principles (Part XI)**

---

• **In Relation to Agency *(cont'd)***

—*The Evolution-Transformation Principle*
*(On the Multiple Kinds of Agency)*

- This principle (and the symmetry-asymmetry principle) are both about *the agency of change*. The word "agency," in a formal definition, refers to "a person or thing through which power is exerted or an end is achieved." (MWD 2007b) It therefore does not have to necessarily involve an intelligent lifeform.
- Because of this dual meaning in agency, the evolution-transformation principle is more concerned with *the multiple kinds of agency*, that is, both about the *evolution* in the state of nature (e.g., an object of natural beauty) and the *transformation* in the world of intelligent lifeforms (e.g., a work of art, an air-conditioner, etc.).
- And the transformative part of the principle precisely refers to the other dimension in the dual meaning of agency, in giving technology a major role to play in the change of the world, which is something that I extensively analyzed in *FHC* in the context of the technophilic lifeworld, espeically though not exclusively since modern times.
- Family resemblance: e.g., inorganicness-volition (or unwillingness-willingness).
- Sources: Especially from *FAE*. See also FHC, *FCD*, *FPHC*, *BDPD*, *BCPC*, *FPHST*, *BCIV*, *BNN*, *BWT*, *FC*, *FIA*, *ALD*, *FPHU*, *FPHK*, *FPHML*, *FPHE*, *FPHMM*, *FPHCT*, *FPHG*, & *FPHUP*.

---

(*continued on next page*)

**Table 4.4. The Syntax of Existential Dialectics I: The Principles (Part XII)**

- **In Relation to Agency** ***(cont'd)***

  —*The Convention-Novelty Principle*
  *(On the Nature of Creative Thinking)*

  - If there is conventional wisdom, there is likewise novel challenge, to the extent that both convergent and divergent thinking are part of life.
  - In addition, in this context of the convention-novelty principle, there are, in the absence of better words, what I want to call (a) *creative techniques* and (b) *creative traits*, which, when satisfied—in relation to the larger context of the mind, nature, society, and culture—can be used to enhance creative works. The list of creative techniques and traits (as summarized in *Table 4.1* of *FPHCT*) are not exhaustive, of course, but the examples here constitute a great beginning of understanding the structure of creative thinking.
  - Yet, lest the reader gets carried away by the euphoria about creative thinking in our time, it should be reminded that creative thinking has its own possibilities and limits (as shown in *Table 2.1* of *FPHCT* on invention), just as it has its own promises and pitfalls (as shown in *Table 3.1* of *FPHCT* on innovation). In the end, it should also be stressed that creative thinking has its own desirability and dark sides (as shown in *Table 4.2* of *FPHCT* ).
  - This has important daily implications for an average individual, and the word "average" here allows different degrees of variation from one individual to another in a population, at any given point in history.
  - Family resemblance: e.g., convergence-divergence
    - Sources: First from *FPHCT*. See also *FPHG*, & *FPHUP*.

(*continued on next page*)

**Table 4.4. The Syntax of Existential Dialectics I: The Principles (Part XIII)**

---

• **In Relation to Outcome**

—*The Regression-Progression Principle*
*(On the Direction of History)*

- Neither the cyclical nor the linear views are adequate for explaining many phenomena at all levels.
- History progresses to more advanced forms, but with a regressive touch. Examples include no freedom without unfreedom, no equality without inequality, and no civilization without barbarity.
- This is not an inevitable law, but merely a highly likely empirical trend.
- Family resemblance: e.g., decline-growth
- Sources: From all my books, starting with *FHC*. First named in *BCPC*.

—*The Sameness-Difference Principle*
*(On the Metamorphosis of Change)*

- An entity, as it evolves over time, can be both different from and similar to its opposing alternatives and does not have to be solely more different from them over time.
- Opposites are not absolute in a black-or-white fashion; so, an entity can become relatively more similar to (or more different from) its opposite over time.
- Family resemblance: e.g., homogeneity-heterogeneity
- Sources: Especially from ALD. See also *FHC, FCD, FPHC, BCPC, BDPD, FPHST, BCIV, BNN, BWT, FC, FAE, FIA, FPHU, FPHK, FPHML, FPHE, FPHMM, FPHCT, FPHG, & FPHUP*.

---

*Notes*: The features in each principle are solely illustrative (not exhaustive). The comparison is also relative (not absolute), nor are they mutually exclusive. As generalities, they allow exceptions.

**Table 4.5. The Syntax of Existential Dialectics II: The Principles as Short Cuts (Part I)**

—The principles should be treated with caution, lest misunderstanding occurs, since they do not constitute rigid dualities (or dichotomies).

—The reason is that each pair in an ontological principle consists of two opposites, which are, however, merely short cuts both for multiple variations and degrees, as well as for different interactions with multiple other entities unlike them. In this light, each pair can end up having hundred (or even thousand, if not more) different versions, which interact with hundred (or even thousand, if not more) other entities. There are two clarifications here.

—Firstly, this conception of shortcuts is not mutually exclusive nor absolute, in that the opposites can come in all shapes and sizes, with different degrees. For instance, by analogy, just as there are different degrees of the two colors "white" and "black"—there are likewise different degrees of the opposites in each ontological pair, to the extent that there can be multiple entities (not only two) interacting with each other in each pair.

—And secondly, this conception of shortcuts do not ignore other possible entities in interacting with the two opposites (with their different versions) in each pair. For instance, by analogy, there are not only the two colors "white" and "black" as opposites, since there are other colors too besides them like "yellow," "green," "purple," or else. The same logic can be applied to each ontological pair, in that they also interact with other entities, not with only two of them (with their different versions).

—Therefore, with these two clarifications in mind—each pair in an ontological principle serves only as an abbreviation for something more complicated and, therefore, although it contains two opposites, it should not be confused as a duality (dualism).

*(continued on next page)*

**Table 4.5. The Syntax of Existential Dialectics II: The Principles as Short Cuts (Part II)**

—Existential dialectics rejects any dualism (or dichotomy) as too rigid and instead allows the multiplicity of entities, to the extent that between the two opposites in each pair exist many other alternatives to choose from. They are named in that short form for aesthetic elegance, instead of listing all possible entities between the two opposites in the title.

—To be dialectic is to go beyond any rigid dichotomy and transcend into something different altogether in the long haul. One may be tempted to call the dialectic logic here with a different name like existential "multilectics" (instead of "dialectics"), but this naming is incorrect (or even misleading), for the two reasons aforestated.

—After all, each ontological pair come in all shapes and sizes (with different degrees) and do not exist by themselves but also interact with other entities unlike them. The virtue of revealing an ontological pair is to show how they relate within themselves (in multiple versions) and also interact with others unlike them (also in multiple versions).

—This will be clear in the table on "the dialectic constraints imposed by the principles."

*Source*: From Ch.1 of *FPHK*

**Table 4.6. The Syntax of Existential Dialectics III: The Principles as Family Resemblances**

—Each ontological principle is generic, with some other comparable ontological pairs to be put in the same family (like a *family resemblance*).
—Ludwig Wittgenstein (1953) once suggested the idea of "family resemblance" in explaining different games classified under the same family called *games*.
—Why should, for instance, playing football and chess as playing "games," when it is well understood that football is not the same as chess? The answer is that, although each game is different and has different rules, many of them (though not all) share, more or less, some commonalites (e.g., scoring as necessary for winning).
—And this is so, even though some games share more than some others in any given selection of critieria, and no two games are exactly identical. So, his point here is that there is no essential core which is common to all games, and the best that one can look for is some characteristics which are common to many (but not all) games. (A. Biletzki 2006)
—By the same logic—in the previous section on selection criteria, the flexibility-inflexibility pair can be put in the family resemblance of the simpleness-complicatedness principle, although the two pairs are not exactly identical. Likewise, the directness-indirectness pair can be put in the family resemblance of the preciseness-vagueness principle, although, again, the two pairs are not exactly identical.
—In this sense, which specific pair in a family should be used to designate the name of the family can be at times a bit arbitrary, but with good reason.
—For illustration, in the context of method, the partiality-totality principle can take the different form like individualisticness-holisticness, just as the explicability-inexplicability principle can take the different form like underlyingness-regularness—although each two pairs are not exactly identical and have slightly different meanings and usages.
—A more comprehensive listing of this family resemblance for all other principles is shown in the table on the syntax of existential dialectics.

*Source*: From Ch.1 of *FPHK*

**Table 4.7. The Syntax of Existential Dialectics IV: The Dialectic Constraints Imposed by the Principles (Part I)**

• **Co-Existent and Asymmetric**

—The principles, as they constitute the syntax of existential dialectics, are dialectic in character, such that, when they are applied, they impose dialectic constraints on how reality is to be understood. Consider, say, the symmetry-asymmetry principle as an illustration here, in order to summarize two main characters of the dialectic constraints in question.

—Firstly, to be dialectic here is to go beyond the narrow dichotomies (and, for that matter, any rigid multi-dimensional classificatory scheme), be they about "self" vs. "world," "freedom" vs. "unfreedom," "barbarity" vs. "civilization," "individuality" vs. "communality," and so on.

—One way to do so (to go beyond) is to consider them all in terms of co-existence (without favoring one over the rest). For instance, my theory of "post-civilization" (to be summarized later in the section on the pragmatics of existential dialectics) is to go beyond barbarity and civilization in terms of understanding barbarity and civilization as being co-existent. And the same logic can be said in relation to my theories of "post-democracy," "post-capitalsm," and others (also to be introduced later in the section on the pragmatics of existential dialectics), in regard to freedom vs. unfreedom, equality vs. inequality, communality vs. individuality, spirituality vs. materiality, and so on.

—But to consider them all (in the dichotomies—and, for that matter, in any rigid multi-dimensional classifactory schme) as co-existent is not the same as to imply that the opposites in any classificatory scheme are all equal, since, in accordance to the symmetry-asymmetry principle (as an illustration here), if they are equal in terms of being considered as co-existent, they are asymmetric in terms of being unequal in dialectic interaction (e.g., X can be more dominant than Y in case A, or Y is more dominant than X in case B).

*(continued on next page)*

**Table 4.7. The Syntax of Existential Dialectics IV: The Dialectic Constraints Imposed by the Principles (Part II)**

—For this reason, there are different versions of "post-democracy" and "post-capitalism" in my theories. As an illustration, in version I of the theory of post-democracy, freedom is more dominant than equality, whereas in version II of the theory of post-democracy, equality is more so than freedom.

—But this "X more than Y" has to be understood in the context of dialectic logic (not in conventional logic), in that both "X" and "Y" are important in post-democracy (in the context of dialectic logic), but in an asymmetry way. By contrast, in conventional logic, it often favors one over the other—be it in regard to privileging freedom over equality in Fascism, favoring freedom relatively more than equality in Liberal Democracy, or favoring equality relatively more than freedom in Socialist Democracy. In the latter two cases (about Liberal Democracy and Socialist Democracy), the difference between dialectic logic and conventional logic can be one in degree, not in kind—in this sense, albeit not in other senses.

—The same logic can be said about the relatonships between individuality and communality, between spirituality and materiality, and between formal legalism and informal legalism in the different versions of my theory of post-capitalism.

- **Transcendent**

—Secondly, to be dialectic is to go beyond the narrow dichotomies (and, for that matter, any rigid multi-dimensional classificatory scheme) in another way, this time, in a transcendent way, that is, in exploring other possibilities or even other issues not considered within the narrow confines of narrow dichotomies (and, for that matter, any rigid multi-dimensional classificatory scheme).

*(continued on next page)*

**Table 4.7. The Syntax of Existential Dialectics IV: The Dialectic Constraints Imposed by the Principles (Part III)**

—As an analogy, to go beyond the narrow color dichotomy of "black" and "white" is not just to choose both "black" and "white" (as in the first meaning) but also to explore other color options (e.g., "green," "purple," "blue," etc.—and, alternatively, "shade," "line," "curve," etc.). By the same logic, to go beyond "democracy" is to transcend democracy (as in version III of the theory of "post-democracy") and to explore other possibilties of lifeforms (e.g., floating consciousness, hyper-spatial consciousness, etc., to live beyond the narrow obsession with freedom and equality).

—This dialectic character of the principles in existential dialectics has important implications for the pragmatics of existential dialectics (as will be clear shortly, in the section on the pragmatics of existential dialectics).

*Notes*: The examples here are solely illustrative (not exhaustive). The comparison is also relative (not absolute), nor are they mutually exclusive. As generalities, they allow exceptions.
*Sources*: First explicitly stated in *ALD*. Also from all other books of mine.

**Table 4.8. The Syntax of Existential Dialectics V: Further Clarifications (Part I)**

- **Pioneering**
  —Firstly, the total number of ontological principles is unknown, to be discovered later, as our knowledge of the world becomes more advanced.
  —Consequently, the principles as introduced in my books are not exhaustive, with new ones being added, whenever more of them are discovered in later research.
  —At least, future generations can pick up where I leave off and continue the discovery.
  —In this sense, my work should be treated as a pioneering effort for the development of a systematic, comprehensive analysis of a new general ontology for the future of knowledge.

- **Flexible**
  —Secondly, the principles are not rigidly classified, as they can be reclassified in a different way.
  —For instance, the preciseness-vagueness principle is classified under the category of "structure" but can be reclassified under the category of "method," although in so doing, it has a different meaning in the context of method.
  —The same logic applies to the same-difference principle under the category of "outcome," which can be reclassified under the category of "structure," although in so doing, once more, it has a different meaning in the context of structure.

- **Mutually Constraining**
  —Thirdly, the principles are to be understood together, not that each prinicple is to be analyzed independently of others. After all, the principles are mutually constraining, in that they work together as a whole.

*(continued on next page)*

**Table 4.8. The Syntax of Existential Dialectics V: Further Clarifications (Part II)**

- **Selectively Useful**
  —Fourthly, the principles are relevant to all subject matters, but some principles are more useful to some subject matters than others—as implied in the symmetry-asymmetry principle. This is true, even if different studies of the same kind can yield different views about the degree of relevance for each ontological principle, depending on the specific nature of a research in question, needless to say.

- **Anti-Reductionistic**
  —And fifthly, the principles constitute only two levels of analysis, this time, at the ontological and methodological levels—while other levels of analysis (from the perspectives of the mind, nature, society, and culture) are also needed, in order to understand reality in its totality (as explained in the section on "sophisticated methodological holism," so as to avoid the dual dangers of reductionism and reverse-reductionism).

*Notes*: The examples here are solely illustrative (not exhaustive). The comparison is also relative (not absolute), nor are they mutually exclusive. As generalities, they allow exceptions.
*Sources*: From my previous books like *FHC*, *FCD*, *FPHC*, *BDPD*, *BCPC*, *BNN*, *FC*, *FAE*, and so on.

**Table 4.9. The Syntax of Existential Dialectics VI: The Dilemma of Specific vs. General Ontology (Part I)**

• Ontology is often more complicated and imposes some seemingly insurmountable difficulties concerning what constitutes a kind of ontology which can be valid enough to be accepted by the wider intellectual community in question. An excellent example concerns what I want to call *the dilemma of ontology* in relation to the relationship between "specific ontology" and "general ontology." (M. Bunge 1999; R. Corazzon 2007)

• On the one hand, "general ontology," as Mario Bunge (1999) put it, "studies all existents." But, on the other hand, "special ontology studies one genus of thing or process—physical, chemical, biological, social, etc." (M. Bunge 1999; R. Corazzon 2007)

• Consequently, the relationship between the former and the latter is that, as an illustration, "whereas general ontology studies the concepts of space, time, and event, the ontology of the social investigates such general sociological concepts as those of social system, social structure, and social change." (M. Bunge 1999; R. Corazzon 2007)

• With this formal definition in mind, the relationship between the two forms of ontology only poses *the dilemma of ontology*, in that general ontology is possible to the extent that its general contructs must be broad enough to serve as the "lowest common denominator" for all disciplines, but this character of the lowest common denominator is not only hard to find but also its usefulness is limited. (WK 2008iii)

• But specific ontology fares no better either, in that its specific constructs primarily serve a specific discipline in question, although other disciplines can appropriate them for their own purposes, but in a different context (especially, though not exclusively, when used as a metaphor from one field to another). For instance, the Kantian category of understanding under the heading of "quantity" constitutes a type of specific ontology, which is specific to the discipline of mathematics, although other disciplines can make use of the categories in their own context.

*(continued on next page)*

**Table 4.9. The Syntax of Existential Dialectics VI: The Dilemma of Specific vs. General Ontology (Part II)**

• A solution lies in my proposal of "sophisticated methodological holism," which requires all levels of analysis in any subject matter. See the tables on sophisticated methodological holism, and on reductionism and reverse-reductionism, for more details.

*Source:* From *Sec. 1.4* of *FPHK*

**Table 4.10. The Semantics of Existential Dialectics**

- **Abstract Structure vs. Specific Meanings**
  —The syntax of existential dialectics so understood in terms of ontological principles only gives us the structure of ontology in the world, in an abstract (general) sense. These principles by themselves do not tell us the specific meanings in a given context.
  —In order to grasp the specific meanings of the principles in a given context, it is necessary to study the semantics of existential dialectics. And the analysis of the ontological principles as family resemblances in the section on syntax is only a starting point (and thus overlaps a bit with the semantics of existential dialectics here).
  —With this caveat in mind, the reason that I have often gone in great lengths in my previous books on different subjects is to explain the specific meanings of the principles when applied in different contexts.

- **Specific Meanings in Specific Fields**
  —For instance, in *FPHST*, I used the first three principles (i.e., the change-constancy principle, the regression-progression principle, and the symmetry-asymmetry principle) to propose "the perspectival theory of space-time," for a better way to understand space and time—especially, though not exclusively, in relation to future post-human history (as summarized in *Table 3.6*, *Table 3.7*, *Table 3.8*, and *Table 3.9*). In so doing, I had to introduce concepts and theories specific to the field of physics and other related fields (e.g., "absolute space" and "absolute time" in "classical mechanics" and "relative space-time" in "the theory of relativity").
  —In *BNN*, I also exploited the three principles to propose the "transcendent" approach to the study of genes and memes as a new way to understand the interaction between nature and nurture. In so doing, I had to explore concepts and theories in the world of evolutionary theory (e.g., "mutation," "variation," "adaptation," "selection," and "inheritance" in Darwinian evolutionary theory) and neural biology (e.g., "chromosome," "gene," "DNA," "RNA," "protein," "neuron," "neural network," and "behavior").

*Notes*: The categories and examples in each are solely illustrative (not exhaustive). The comparison is also relative (not absolute), nor are they mutually exclusive. As generalities, they allow exceptions.
*Sources*: First explicitly stated in *ALD*. Also from all other books of mine.

**Table 4.11. The Pragmatics of Existential Dialectics (Part I)**

- **The Two-Way Street Connecting Theory and Meta-Theory**
  —The pragmatics of existential dialectics is not a one-way street (that is, using the ontological principles for theoretical insights in praxis) but a two-way one, that is, (a) from meta-theory to theory, and (b) from theory to meta-theory.
  —(a) On one side of the street, the ontological principles can inspire some theoretical insights in praxis, that is, in relation to some specific fields.
  —(b) On the other (opposing) side of the street, however, the study of a subject matter in the specific fields in turn reveals some more hitherto unknown ontological principles to be discovered and identified. For this reason, three new principles were added in *FC*, one in *FAE*, and two in *ALD*, on top of the original three in *BCPC*—after some research on the specific subject matters.

- **Direct and Indirect Applications**
  —*Direct*
    - The logic of existential dialectics can shed some theoretical insights on diverse phenomena in the world, and good instances are the pertinent use of the principles of existential dialectics for the theoretical insights on the freedom/unfreedom dialectics, the equality/inequality dialectics, and the wealth/poverty dialectics in my previous works.
    - My latest books like *FPHST* and *BNN* also use the principles to reveal some theoretical insights on the perspectives of space and time (as in *FPHST*) and of nature and nurture (as in *BNN*).

  —*Indirect*
    - The theoretical insights can further be used to reveal other phenomena directly from them (viz., the theoretical insights) and therefore indirectly from the principles themselves. A good instance is the use of the theoretical insights on the freedom/unfreedom and equality/inequality dialectics for the understanding of the civilization/barbarity dialectics.

*(continued on next page)*

**Table 4.11. The Pragmatics of Existential Dialectics (Part II)**

---

- **Direct and Indirect Applications (*cont'd*)**
  - Even in indirect applications, however, a phenomenon under study can still be directly related back to the principles themselves. In the example as cited above, the civilization/barbarity dialectics can be directly related to the principles of existential dialectics without the intermediate role of the freedom/unfreedom and equality/inequality dialectics.

- **Multiple Levels of Application**

  —The theoretical insights can be applied to different levels of analysis, even though in a given example, it may refer to one level only. For instance, in the example concerning the freedom/unfreedom dialectics, it can be used at the structural level (e.g., in relation to the theory of cyclical progression of hegemony), but it can be exploited as well for other levels (e.g., the theory of post-capitalism at the institutional level).

---

*Notes*: The categories and examples in each are solely illustrative (not exhaustive). The comparison is also relative (not absolute), nor are they mutually exclusive. As generalities, they allow exceptions.

*Sources*: From Ch.6 of *BCPC*. See also *FHC*, *FCD*, *FPHC*, *BDPD*, *BNN*, *FC*, *FAE*, *ALD*, *FIA*, *FPHU*, *FPHK*, *FPHML*, *FPHE*, *FPHMM*, *FPHCT*, & *FPHG*.

**Table 4.12. The Freedom-Unfreedom Dialectics (Part I)**

- **On Having**

—*In Relation to the Technological*

(1) if freer from submission to Nature, then less free from ecological degradation (Deep and Social Ecology), even if in a hi-tech form

(2) if freer from technological inconvenience / backwardness, then less free from technological control and the loss of privacy

(3) if freer from technological (material) backwardness, then less free from the abusive (barbaric) maltreatment of the primitive Others

—*In Relation to the Everyday*

(1) if freer from abject poverty, then less free from artificial needs/discontents (Frankfurt School)

(2) if freer from sensual suppression, then less free from violent sublimation (Freud)

(3) if freer from the snobbishness of high culture, then less free from the shabbiness (leveling-off effect) of mass culture (Tocqueville)

(4) if freer from the inefficiency of traditional "compassionate economy," then less free from the bondage of a "ruthless [competitive] economy" (Keynes)

(5) if freer from anarchy in the state of nature (system fragmentation), then less free from government regulations and controls in system integration

*(continued on next page)*

**Table 4.12. The Freedom-Unfreedom Dialectics (Part II)**

- **On Belonging**
  - —*In Relation to the Good and the Just*
    - (1) if freer from disciplinary society, then less free from society of control (Foucault)
    - (2) if freer from the tyranny of one or a few, then less free from the tyranny of the majority (or sometimes, minority veto)
    - (3) if freer from elitist decision making, then less free from political gridlock/cleavage
    - (4) if freer from arbitrary (discretionary) administration, then less free from bureaucratic irrationality (Weber) and legal trickery (loopholes)

- **On Being**
  - —*In Relation to the True*
    - (1) if freer from unscientific dogmas, then less free from instrumental abyss (nihilism). Or conversely, if freer from meaninglessness, then less free from dogmas.
    - (2) if freer from the bondage of partiality/partisanship (e.g., prejudice, discrimination), then less free from the danger of impartiality and neutrality (e.g., opportunism, unrealisticness, lack of compassion, inaction)
    - (3) if freer from making generalizations, then less free from being unable to understand much of anything
  - —*In Relation to the Holy*
    - (1) if freer from collective conscience, then less free from social loneliness
    - (2) if freer from religious absoluteness, then less free from spiritual emptiness
  - —*In Relation to the Beautiful/Sublime*
    - (1) if freer from artistic non-autonomy, then less free from aesthetic disillusion (deconstruction)

*(continued on next page)*

**Table 4.12. The Freedom-Unfreedom Dialectics (Part III)**

*Notes*: The examples in each category are solely illustrative (not exhaustive), and the comparison is relative (not absolute), nor are they necessarily mutually exclusive. And some can be easily re-classified elsewhere. As generalities, they allow exceptions.
*Sources*: A reconstruction from Ch.10 of *FCD*, based on *FHC*

**Table 4.13. The Equality-Inequality Dialectics (Part I)**

- **On Having**
  —*In Relation to the Technological*
  (1) if more equal in treating Nature with spiritual unity, then less equal in suppressing the dominant drive to transcend it altogether
  —*In Relation to the Everyday*
  (1) if more equal in building social plurality, then less equal in leveling-off effects (e.g., the subsequent relative intolerance of high/intellectual ethos in mass culture industry)
  (2) if more equal in socioeconomic distribution beyond a certain point, then less equal in efficiency (e.g. resentment, the erosion of work ethics)
  (3) if more equal in urging an affirmative action program, then less equal in creating victim mentality (in oneself), stigma (from others), reverse discrimination (against the once privileged), and mediocracy (against the more able)

- **On Belonging**
  —*In Relation to the Good and the Just*
  (1) if more equal in banning monarchic/oligarchic exclusion, then less equal in producing "the tyranny of the majority" or of "minority veto"
  (2) if more equal in encouraging participatory decision making, then less equal in inducing political divisiveness (gridlock/cleavage in power blocs) and organizational oligarchy
  (3) if more equal in institutionalizing a decentralized bureaucracy, then less equal in falling into more territorial/turf politics (intrigues)

*(continued on next page)*

**Table 4.13. The Equality-Inequality Dialectics (Part II)**

---

- **On Being**
  - —In Relation to the Beautiful / Sublime
    - (1) if more equal in accepting diverse styles ("anything goes" mentality), then less equal in artistic good quality (in leveling-off effects against the best)
  - —In Relation to the True
    - (1) if more equal in tolerating multiple viewpoints (no matter how extreme), then less equal in epistemic standards
  - —In Relation to the Holy
    - (1) if more equal in celebrating any cults and sects (no matter how questionable), then less equal in spiritual depth and authenticity

---

*Notes*: The examples in each category are solely illustrative (not exhaustive), and the comparison is relative (not absolute), nor are they mutually exclusive. And some can be easily reclassified elsewhere. As generalities, they allow exceptions.
*Sources*: A reconstruction from Ch.10 of *FCD*, based on *FHC*

**Table 4.14. The Duality of Oppression in Existential Dialectics: Oppression and Self-Oppression (Part I)**

- **From the Same to the Others and Itself**
  - —The Oppression by the Same against the Others
    - *By way of downgrading differences*
      - –Ex: on judiciary caprice for corporate crimes (*Sec.2.2.1.2.1*)
      - –Ex: on the deceptive politics of liberation (*Sec.3.5*)
      - –Ex: on the humanitarian mystique (*Sec.4.4*)
      - –Ex: on the fad of emotional intelligence (*Sec.5.3*)
    - *By way of accentuating differences*
      - –Ex: on the legal sophistry of self-defense (*Sec.2.3*)
      - –Ex: on the legal semantics of proportionality (*Sec.2.4*)
      - –Ex: on the tricky politics of external threat (*Sec.3.4*)
      - –Ex: on the appeal of the Far Right for democracy (*Sec.5.4*)
      - –Ex: on the democratic axis of evil (*Sec.5.5*)
      - –Ex: on the democratic way of brutality and revenge (*Sec.5.6*)
      - –Ex: on democratic autocracy (*Sec.6.4*)
  - —The Oppression by the Same against Itself
    - *By way of downgrading differences*
      - –Ex: on the politics of fear (*Sec.2.2*)
      - –Ex: on the trickery of compassionate conservatism (*Sec.3.2*)
      - –Ex: on the deceptive politics of patriotism (*Sec.3.3*)
    - *By way of accentuating differences*
      - –Ex: on the caprice of due process on domestic suspects (*Sec.2.2*)
      - –Ex: on the false security/freedom dilemma (*Sec.6.5.2*)

*(continued on next page)*

**Table 4.14. The Duality of Oppression in Existential Dialectics: Oppression and Self-Oppression (Part II)**

- **From the Others to the Same and Themselves**
  - —The Oppression by the Others against the Same
    - *By way of downgrading differences*
      - –Ex: on judiciary caprice in the reverse direction (*Sec.2.2.1.2.2*)
      - –Ex: on equal pay (*Sec.6.2.1.1*)
      - –Ex: on equal representation (*Sec.6.2.1.2*)
      - –Ex: on affirmative action program (*Sec.6.3.1.1*)
      - –Ex: on same-sex marriage (*Sec.6.3.1.2*)
    - *By way of accentuating differences*
      - –Ex: on sexual harassment (*Sec.6.2.2.1*)
      - –Ex: on physical violence (*Sec.6.2.2.2*)
      - –Ex: on sexual exploitation (*Sec.6.2.2.3*)
  - —The Oppression by the Others against Themselves
    - *By way of downgrading differences*
      - –Ex: on the reverse-class mystique (*Sec.4.2*)
      - –Ex: on the reverse-black mystique (*Sec.4.3*)
      - –Ex: on self-discrimination by downgrading (*Sec.6.3.2.2*)
    - *By way of accentuating differences*
      - –Ex: on self-discrimination by accentuating (*Sec.6.3.2.1*)

*Notes*: The examples are solely illustrative (not exhaustive), nor are they mutually exclusive. As generalities, they allow exceptions. Also, both forms of oppression co-exist in all of the examples, so the listing of them are only meant in a relative, not absolute, sense.
*Source*: A summary of the sections (as cited) in Chs.2-6 of *BDPD*. See text for more info and references.

**Table 4.15. The Structure of Existential Dialectics I: The Freedom/Unfreedom and Equality/Inequality Dialectics**

- Each freedom and equality produces its own unfreedom and inequality, regardless of whether the pair occurs in political society (with the nation-state), in civil society (with some autonomy from the state), or elsewhere (e.g., in the private sphere of individual homes)—and regardless of whether freedom and equality are understood as "negative" or "positive."

- Oppression is dualistic, as much by the Same against the Others and itself, as by the Others against the Same and themselves.

- Both forms of oppression and self-oppression can be achieved by way of downgrading differences (between the Same and the Others) and of accentuating them.

- The relationships are relatively asymmetric between the Same and the Others and relatively symmetric within them. This is true, even when the Same can be relatively asymmetric towards itself in self-oppression, just as the Others can be likewise towards themselves.

- Symmetry and asymmetry change over time, with ever new players, new causes, and new forms.

*Notes*: The examples in each category are solely illustrative (not exhaustive) nor necessarily mutually exclusive, and the comparison is relative (not absolute). As generalities, they allow exceptions. "Negative" freedom is freedom "from" (e.g., freedom from poverty), whereas "positive" freedom is freedom "to" (e.g., freedom to the state of enlightenment). "Negative" equality is "procedural" equality (e.g., equality of opportunity), while "positive" equality is "substantive" equality (e.g., equality of outcome). Existential dialectics impose constraints on freedom and equality in democracy, non-democracy, and post-democracy. There is no utopia, in the end; even should there be one, dystopia would exist within it.
*Sources*: From *Table 1.5* of *BDPD*—and also from *FHC*, *FCD*, and *FPHC*

**Table 4.16. The Structure of Existential Dialectics II: The Wealth/Poverty Dialectics**

• There is no wealth without poverty, just as there is no poverty without wealth.

• The wealth/poverty dialectics occurs in the realms of having, belonging, and being, in relation to the material, relational, and spiritual.

• The wealth/poverty dialectics also expresses itself at the multiple levels of analysis in accordance to methodological holism, be they about the micro-physical, the chemical, the biological, the psychological, the organizational, the institutional, the structural, the systemic, the cultural, and the cosmological.

• The wealth/poverty dialectics is a different manifestation of existential dialectics in general, subject to the principles in its logic of ontology—just as the freedom/unfreedom and equality/inequality dialectics are likewise.

• There is no economic utopia, in the end; even should there be one, dystopia would exist within it.

*Notes*: The main points here are solely illustrative (not exhaustive) nor necessarily mutually exclusive, and the comparison is relative (not absolute). As generalities, they allow exceptions.
Sources: From *BCPC*. See also *FCD* and *FHC*.

**Table 4.17. The Structure of Existential Dialectics III: The Civilization/Barbarity Dialectics**

• There is no civilization without barbarity.

• The civilization/barbarity dialectics applies in the four civilizing processes (e.g., the rationalizing process, the pacifying process, the stewardizing process, and the subliming process).

• The civilization/barbarity dialectics is another (different) manifestation of existential dialectics in general, subject to the principles in its logic of ontology—just as the freedom/unfreedom and equality/inequality dialectics and the wealth/poverty dialectics are likewise.

• There is no utopia, in the end; even should there be one, dystopia would exist within it.

*Notes*: The main points here are solely illustrative (not exhaustive) nor necessarily mutually exclusive, and the comparison is relative (not absolute). As generalities, they allow exceptions.
Sources: From *BCIV*. See also *FCD*, *FHC*, and *BDPD*.

**Table 4.18. The World of Titans, and the Remaking of the World Order**

- **Hyper-Empires**
  - —Ex: The Chinese Union
  - —Ex: The Indian Union

- **Meso-Empires**
  - —Ex: The European Union
  - —Ex: The North American Union

- **Micro-Empires**
  - —Ex: The Latin American Union
  - —Ex: The Middle Eastern Union

- **The Rest of the World**
  - —<u>Odd Powers</u>
    - Ex: Japan
    - Ex: Russia
  - —<u>The Poor Club</u>
    - Ex: The African Union
  - —<u>Ambivalent Regions</u>
    - Ex: Southeast Asia
    - Ex: Oceania
    - Ex: South Asia
    - Ex: Central Asia
    - Ex: Southern/Eastern Europe
    - Ex: North Africa
    - Ex: Central America
    - Ex: Others (e.g., the Korean peninsula)

*Source*: A summary of Chs.2-5 (of *BWT*)

**Table 4.19. The Origins of Authoritarian Liberal Democracy**

- **The Geopower of Nature (Ch.4 of *ALD*)**
  —Ex: Power Character and Geographical Strategy
  —Ex: Living Space and Territorial Expansion
  —Ex: Strategic Heartland and Containment

- **The Biopsychology of the Mind (Ch.5 of *ALD*)**
  —Ex: The Bell Curve and Mass Intelligence
  —Ex: Group Analysis and Mass Knowledge
  —Ex: Groupthink and Elite (Mis)calculation

- **The Disciplinary Control of Society (Ch.6 of *ALD*)**
  —Ex: Social Organizations of Ruthlessness
  —Ex: Social Institutions of Greed
  —Ex: Social Structure of Exclusion
  —Ex: Social Systems of Violence

- **The Molding Force of Culture (Ch.7 of *ALD*)**
  —Ex: The Tradition of Conquest
  —Ex: The Rationalization of Unreason

*Notes*: The examples in each category are solely illustrative (not exhaustive), and they are also relative (not absolute), nor are they mutually exclusive. As generalities, they allow exceptions.
*Source*: A summary of Chs.4-7 of *ALD*

**Table 4.20. The Theory of Post-Democracy I: The Priority of Freedom over Equality (Part I)**

- **Differences**
  - *—For the aggressive Lions (the strong Elitists)*
    - •Setting up rank distinctions among unequals (e.g., between inferior humans and superior post-humans, or later among inferior post-humans and superior ones, relatively speaking)
    - •Yearning for being not only distinguished from unequals, but also the first among equals (the best of the very best)
    - •Soul-searching for a high spiritual culture (not the trashy one for the masses). Mass culture is a dirty joke for them.
  - *—For the manipulative Foxes (the weak Counter-Elitists)*
    - •Seeking a gentle hegemony by way of more communitarian concerns ( for inferior humans and, later, inferior post-humans)
    - •Being more sympathetic to less formal-legalistic institutions and values

- **Similarities**
  - *—For both Lions and Foxes*
    - •Exploring different spheres of non-human consciousness in the cosmos (something vastly superior than the human one)
    - •Recognizing the democratic illusions (e.g., no freedom without unfreedom, no equality without inequality, or simply no justice without injustice, and vice versa)

*(continued on next page)*

**Table 4.20. The Theory of Post-Democracy I: The Priority of Freedom over Equality (Part II)**

*Notes*: The two callings and examples in each category are solely illustrative (not exhaustive), since there will be many different post-human value ideals in the distant future of post-human civilization. The comparison is also relative (not absolute) towards post-democracy, so this is not just a version of free-market democracy (nor Fascism/Nazism, as shown in the table later on democracy, non-democracy, and post-democracy). Nor are they mutually exclusive. As generalities, they allow exceptions. And the specific forms of post-human post-democratic ideals need to be further developed in future after-postmodern history, as they will be different from the ones we now know. The point here is to solely give an extremely rough picture of a small part of the world to come that we have never known.
*Source*: From Ch.10 of *FCD*. Refer to text for more info and references.

**Table 4.21. The Theory of Post-Democracy II: The Priority of Equality over Freedom**

- **Hybrid Versions of**
  —Ex: the Trans-Feminine Calling
  —Ex: the Trans-Sinitic Calling
  —Ex: the Trans-Islamic Calling
  —Ex: the Trans-Outerspace Calling

- **Qualifications**
  —These four versions of post-capitalist value ideals need not automatically be post-democratic, just as capitalism does not necessarily mean democracy. They are two different entities—though closely related.
  —But up to a certain threshold of elevating equality at the farther expense of freedom, the democratic ideals will be overcome and cease to exist.
  —The overcome will not be socialist or communist, but post-democratic with no freedom without unfreedom and no equality without inequality, subject to the constraints of existential dialectics.

*Notes*: The callings are solely illustrative (not exhaustive), since there will be many different post-human value ideals in the distant future of post-human lifeforms. The comparison is also relative (not absolute), nor are they mutually exclusive. As generalities, they allow exceptions. And the specific forms of post-human post-democratic ideals need to be further developed in future after-postmodern history, as they will be different from the ones we now know. The point here is to solely give an extremely rough picture of a small part of the world to come that we have never known.
*Source*: From Ch.10 of *FCD*. Refer to text for more info and references.

**Table 4.22. The Theory of Post-Democracy III: The Transcendence of Freedom and Equality (Part I)**

- **Transcending Freedom in Floating Existence**
  —*Freedom*: seeking an ultimate elimination of the body. Being without the body. The aim is to transcend freedom in the end into a metaphysical state (i.e., beyond the physique).
  —*Unfreedom*: yet facing difficult trade-offs. The sacrifice of bodily existence and its joyfulness. An eternal boredom in floating existence in dark deep space, though with alternative pleasures. There is no free lunch even in the state of transcending freedom.

- **Transcending Equality in the Rivalry of Cosmic Hegemony**
  —*Inequality*: competing to outlast other lifeforms in floating existence, or just marginalizing them for one's hegemonic expansiveness in the rest of the cosmos (and even beyond). Universalism is only for the mediocre.
  —*Equality*: accepting only those of one's rank as equal partners in the vast spacetime for cosmic supremacy. Even here, the aim is to transcend equality into a metaphysical state.

*(continued on next page)*

**Table 4.22. The Theory of Post-Democracy III: The Transcendence of Freedom and Equality (Part II)**

*Notes*: Do not confuse this transcendence of freedom and equality (as one version of post-democracy) with the naïve temptation to transcend the freedom/unfreedom and equality/inequality dialectics. Existential dialectics hold true for freedom and equality in all cultures and societies—past, present, or future (i.e., democracy, non-democracy, and post-democracy), regardless of whether freedom and equality are conventionally understood as "negative" or "positive."
Also, the two features and examples in each are solely illustrative (not exhaustive), since there will be many different post-human value ideals in the distant future of post-human lifeforms. The comparison is also relative (not absolute), nor are they mutually exclusive. As generalities, they allow exceptions. And the specific forms of post-human ideals even for these radically alien floating lifeforms (and others unknown to us) need to be further developed in future after-postmodern history, as they will likely be different from the ones herein illustrated. The point here is to solely give a very rough picture of a small part of the extremely alien world to come that we have never known.
*Source*: From Ch.10 of *FCD*. Refer to text for more info and references.

**Table 4.23. Democracy, Non-Democracy, and Post-Democracy (Part I)**

---

• **Democracy**

—*Theoretical Constructs*

•The pursuit of freedom and equality (in various degrees), regardless of whether freedom and equality can be understood as "negative" or "positive"

(1) more equality than freedom: The relative priority of the good over the right

(2) more freedom than equality: The relative priority of the right over the good

—*Types*

•Only (1): Different versions of communitarian moral universalism

•Only (2): Different versions of liberal moral universalism

•(1) or (2): Different versions of anarchic (non-nation-state) moral universalism

•(1) or (2): Different versions of postmodern moral localism

• **Non-Democracy**

—*Theoretical Constructs*

•The focus on (1') equality or (2') freedom, but not both, regardless of whether freedom and equality can be understood as "negative" or "positive"

—*Types*

•Only (1'): Different versions on the Far Left (e.g., Stalinism, Robespierrianism)

•Only (2'): Different versions on the Far Right (e.g., Nazism, absolute monarchism)

---

*(continued on next page)*

**Table 4.23. Democracy, Non-Democracy, and Post-Democracy (Part II)**

- **Post-Democracy**
  —*Theoretical Constructs*
  - The priority of (1'') equality over freedom, or (2'') freedom over equality, or (3'') the transcendence of freedom and equality, regardless of whether freedom and equality are "negative" or "positive." In degree, (1'') or (2'') is less than (1') or (2') but more than (1) or (2)—respectively.
  - Like democracy and non-democracy, post-democracy is also subject to the freedom/unfreedom and equality/inequality dialectics (or existential dialectics in general). Unlike them, post-democracy acknowledges the constraints of existential dialectics and no longer value freedom and equality as sacred virtues. There is no utopia, in the end; even were there one, dystopia would exist within it.

  —*Types*
  - (1''): Different versions of trans-Sinitic value ideals
  - (1''): Different versions of trans-feminine value ideals
  - (1''): Different versions of trans-Islamic value ideals
  - (1''): Different versions of trans-outerspace value ideals
  - (2''): Different versions of post-human elitist value ideals
  - (3''): Different versions of the value ideals of floating consciousness (etc.)

*(continued on next page)*

**Table 4.23. Democracy, Non-Democracy, and Post-Democracy (Part III)**

*Notes*: The examples are solely illustrative (not exhaustive), nor are they mutually exclusive. As generalities, they allow exceptions. “Negative” freedom is freedom “from” (e.g., freedom from poverty), whereas “positive” freedom is freedom “to” (e.g., freedom to the state of enlightenment). “Negative” equality is “procedural” equality (e.g., equality of opportunity), while “positive” equality is “substantive” equality (e.g., equality of outcome). Existential dialectics impose constraints on freedom and equality in democracy, non-democracy, and post-democracy, regardless of whether freedom and equality can be understood as “negative” or “positive” in conventional discourse. Therefore, do not confuse the transcendence of freedom and equality in (3’’) with the naïve temptation to transcend existential dialectics. There is no utopia, in the end; even should there be one, it would not exist without dystopia embedded within it.
*Sources*: A summary, based on my previous works, especially Ch.5 of *FHC*, Chs.5-10 of *FCD*, Chs.2-4 of *FPHC*, and Chs.1 & 7 of *BDPD*. The reader should consult the books for more analysis, as this is only a summary here.

**Table 4.24. Multiple Causes of the Emergence of Post-Democracy (Part I)**

- **At the Micro-Physical Level**
  —Ex: intelligent life without the human physical-chemical system
  —Sources: Ch.7 of *FHC*; Chs.9-10 of *FCD*; Ch.1 of *FPHC*

- **At the Chemical Level**
  —Ex: space radiation and toxins
  —Sources: Ch.7 of *FHC*; Chs.9-10 of *FCD*

- **At the Bio-Psychological Level**
  —Ex: exo-biological evolution in deep space
  —Ex: genetic engineering of new beings
  —Ex: limits of cognitive partiality
  —Ex: illusions of emotional neutrality
  —Ex: human biological inequality
  —Ex: the rise of unfolding unconsciousness
  —Sources: Ch.2 & Chs.9-10 of *FCD*; Ch.7 of *FHC*; Ch.4 of *BCPC*; *FPHU*

- **At the Institutional Level**
  —Ex: the flawed logic of equality
  —Ex: the conflicting nature of governance
  —Sources: Ch. 5 of *FHC*; Chs. 6 & 10 of *FCD*; Ch. 3 of *FPHC*; Chs.2-5 of *BDPD*

- **At the Organizational Level**
  —Ex: e-civic alienation
  —Ex: the dark sides of formal-legalistic routines
  —Sources: Ch.3 of *FHC*; Ch.7 of *FCD*; Ch.3 of *FPHC*

- **At the Structural Level**
  —Ex: ever new forms of inequities, at home and abroad
  —Ex: the emergence of China, women, and Islam as major actors
  —Sources: Chs.5-6 of *FHC*; Chs.7, 9 & 10 of *FCD*; Chs.4-5 of *BDPD*

*(continued on next page)*

**Table 4.24. Multiple Causes of the Emergence of Post-Democracy (Part II)**

---

- **At the Cultural Level**
  —Ex: freedom/unfreedom dialectics
  —Ex: equality/inequality dialectics
  —Ex: system fragmentation and integration
  —Sources: Ch.5 of *FHC*; Chs. 3, 9 & 10 of *FCD*; Ch.4 of *FPHC*; Ch.1 of *BDPD*; Ch.4 of *BCPC*

- **At the Systemic Level**
  —Ex: space habitats (in zero-gravity) and colonization
  —Ex: ultra advanced future info systems
  —Ex: qualitative demography
  —Sources: Ch.7 of *FHC*; Chs.9 &10 of *FCD*

- **At the Cosmological Level**
  —Ex: the colonization of multiverses
  —Ex: the alteration of space-time and the creation of new matter-energy
  —Ex: the expansion of floating consciousness
  —Ex: the spread of hyper-spatial consciousness
  —Sources: Ch.7 of *FHC*; Chs.9 &10 of *FCD*; Ch.4 of *FPHC*; *FPHST*

---

*Notes*: The examples in each category are solely illustrative (not exhaustive), and some of the items can be reclassified somewhere else. Nor are they always mutually exclusive. Since they are generalities, exceptions are expected.
*Sources*: Especially from *FHC*, *FCD*, *FPHC*, *BCPC*, *BDPD*, *FPHST*, and *FPHU*. See also other books and my perspectives on civilizational holism.

**Table 4.25. Some Clarifications about Post-Capitalism and Post-Democracy (Part I)**

• The prefix "trans-" in the first category of post-capitalism (with its four versions) refers to something "going beyond" (not "uniting" or "combining"). Ex: *Sec.10.3.3* of *FCD*; *Sec.2.4* & *Sec.4.4* of *FPHC*; *Sec.7.2* of *BCPC*

• Such terms like "post-democracy," "post-capitalism," "post-human elitist," "trans-feminine calling," and the like as used in my works are more for our current intellectual convenience than to the liking of future humans and post-humans, who will surely invent more tasteful neologisms to call their own eras, entities, and everything else, for that matter. But the didactic point here is to use the terms to foretell what the future might be like, not that its eras and entities must be called so exactly and permanently.
Ex: *Sec.11.1* of *FCD*; *Sec.7.2* of *BCPC*

• The four versions in the first category of post-capitalist value ideals need not automatically be post-democratic, just as capitalism does not necessarily mean democracy. They are two different entities—though closely related. But up to a certain threshold of elevating equality at the farther expense of freedom, the democratic ideals will be overcome and cease to exist. The same is true for the post-human elitist calling in the second category of post-capitalism in relation to post-democracy, depending on the extent to which freedom is elevated at the expense of equality.
Ex: *Sec.10.4.3.3* of *FCD*; *Table 3.9* of *FPHC*; *Table 7.6* of *BDPD*

• The comparison in each of the three realms of existence in all forms of post-capitalism is not absolute, but relative. Examples include "communal" vs. "individualistic," and the like.
Ex: Notes in *Table 10.8*, *Table 10.9*, *Table 10.10*, & *Table 10.11* of *FCD*; Chs.2-4 of *FPHC*; *Sec.7.2* of *BCPC*

*(continued on next page)*

**Table 4.25. Some Clarifications about Post-Capitalism and Post-Democracy (Part II)**

• The emergence of post-capitalism (and post-democracy, for that matter) has multiple causes (to not be reduced to one or only a few).
Ex: Ch.10 of *FCD*, Chs.2-4 of *FPHC*; *Sec.1.3* & *Sec.7.2* of *BCPC* (or *Table 1.8* & *Table 7.11*)

• The specific forms of post-capitalism (and post-democracy, for that matter) need to be further developed in future after-postmodern history, as they will be different from the ones we now know. The point here is to solely give an extremely rough sketch of a world to come that we have never known.
Ex: *Sec.10.3.3* & *Sec.10.4.3.3* of *FCD*; *Table 10.14* & *Table 10.15* of *FCD*; *Sec.7.2* of *BCPC*

• All forms of post-capitalism are not part of a "teleological law," but of "historical trends" only. The same is also true for all forms of post-democracy.
Ex: *Sec.7.1* of *FHC*; *Sec.9.5.3.2* & *Sec.10.3.4.2* of *FCD*; *Sec.7.2* of *BCPC*

• Post-capitalism is not better than capitalism in an "absolute" sense but only fits in better, on the basis of the historical contingency of culture, society, nature, and the mind in some future eras. The same is true for post-democracy in relation to democracy. The term "better" is historically relative.
Ex: *Sec.10.3.3* of *FCD*; *Sec.1.7* of *BDPD*; *Sec.1.5* of *BCPC*

*(continued on next page)*

**Table 4.25. Some Clarifications about Post-Capitalism and Post-Democracy (Part III)**

---

• All forms of post-capitalism and post-democracy are subject to the constraints of existential dialectics. In the process, the dialectic direction is to go beyond the conventional "either-or" dichotomies (e.g., freedom vs. unfreedom, equality vs. inequality, freedom vs. equality, individuality vs. communality, spirituality vs. materiality, formal legalism vs. informal legalism, etc.). As is true in post-civilization, to go beyond the dichotomies is to acknowledge the co-existence of both in each dichotomy, although the degree of scaling one over the over varies from case to case (e.g., the theory of post-capitalism I, the theory of post-capitalism II, the theory of post-democracy I, the theory of post-democracy II, etc.)—but is not to be extreme in largely favoring one over the other, *on average* (all things considered). There is no utopia to be had in the end; even should there be one, dystopia would exist within it.
Ex: Ch.5 of *FHC*; *Sec.10.4.4.2* of *FCD*; *Sec.1.5* of *BDPD*; *Sec.1.3* of *BCPC*; *BCIV*

• All forms of post-capitalism, however different from each other though they are, share one common feature, in that they all inspire for a higher spiritual culture. The same is also true for post-democracy.
Ex: *Sec.10.3*, *Sec.10.4* & *Sec.10.5* of *FCD*; *Chs.2-4* of *FPHC*; *Sec.7.2* of *BCPC*

• All forms of post-capitalism try to avoid the excess in capitalist consumerism by favoring more basic than artificial needs in having, but the quality and quantity of these "basic" needs will be measured by future standards, not by our current ones. Standards are historically relative.
Ex: *Sec.10.3*, *Sec.10.4* & *Sec.10.5* of *FCD*; Ch.2 of *FPHC*; *Sec.7.2* of *BCPC*

---

*(continued on next page)*

**Table 4.25. Some Clarifications about Post-Capitalism and Post-Democracy (Part IV)**

- All forms of post-capitalism make use of a different degree of political authority with advanced info systems in future history and strives for higher spiritual cultures (especially in the post-human age), while acknowledging the constraints of existential dialectics and no longer valuing free market (as in capitalism) and economic control (as in non-capitalism) as sacred virtues.

Ex: *Sec.10.3.4.2*, *Sec.10.3*, *Sec.10.4* & *Sec.10.5* of *FCD*; Chs.2-4 of *FPHC*; *Sec.1.5* of *BDPD*; *Sec.7.2* of *BCPC*

*Notes*:: The main points here are solely illustrative (not exhaustive) nor necessarily mutually exclusive, and the comparison is relative (not absolute). As generalities, they allow exceptions. The sections as cited are only illustrative (not exhaustive).

*Sources*: From *FHC*, *FCD*, *FPHC*, and *BDPD*

**Table 4.26. The Theory of Post-Capitalism I.1: By Group— Ex: Spiritual/Communal in the Trans-Feminine Calling**

- **More Communal Than Individual**
  —*Sharing*: learning from others, as different ideas mutually enrich
  —*Cooperative*: encouraging a sense of shared leadership and teamwork

- **More Informal-Legalistic Than Formal-Legalistic**
  —*Specific*: listening more from the heart than from the head, to know a person as a concrete, not as an abstract, unit
  —*Affective*: thinking and acting with others on a more affective tone. Business can mix with an emotional touch.
  —*Ascriptive*: hiring (or firing) can be done on the basis of merit (or lack of it), but deep solidarity (sisterhood) is important too.
  —*Particularistic*: making decisions on the basis of cost-benefit analysis, but a given group relationship is vital

- **More Spiritual Than Secular**
  —*Long-Term Looking*: sharing for a long-term relationship (e.g., love, friendship), not just for a short-term gain
  —*Loving/Caring*: showing compassion for the sufferings of others, without quickly blaming and pre-judging
  —*Respectful*: showing acceptance about others' feelings (and thoughts)

*Notes*: The categories and examples are solely illustrative, since there can be different versions, and the comparison is relative (not absolute), nor are they mutually exclusive. As generalities, they allow exceptions. The specific forms of the trans-feminine version need to be further developed in future after-postmodern history, as they will be different from the ones we now know, since the prefix "trans-" here means going beyond or deconstructing the feminine values, while using them as the inspirational point at the beginning.
*Source*: From Ch.10 of *FCD*. Refer to text for more info and references.

**Table 4.27. The Theory of Post-Capitalism I.2: By Nation-State— Ex: Spiritual/Communal in the Trans-Sinitic Calling**

- **More Communal Than Individualistic**
  —*Centralized*: being more top-down in management
  —*Collective*: encouraging more group cooperation
  —*Social*: investing in trust and connection

- **More Informal-Legalistic Than Formal-Legalistic**
  —*Specific*: knowing more of those related or connected
  —*Affective*: behaving in a paternalistic, hierarchical way
  —*Ascriptive*: favoring family members and those related
  —*Particularistic*: building connection (guanxi) as imperative

- **More Spiritual Than Secular**
  —*Expansionist*: diffusing civilizational values (e.g., the superiority complex of civilizationalism)
  —*Holistic*: synthesizing things into a panoramic horizon
  —*Historical*: learning from the lessons of the ancient past
  —*Respectful*: deferential to elders and superiors

*Notes*: The categories and examples are solely illustrative, since there can be different versions, and the comparison is relative (not absolute), nor are they mutually exclusive. As generalities, they allow exceptions. The specific forms of the trans-Sinitic version need to be further developed in future after-postmodern history, as they will be different from the ones we now know, since the prefix "trans-" here means going beyond or deconstructing the Sinitic values, while using them as the inspirational point at the beginning.
*Source*: From Ch.10 of *FCD*. Refer to text for more info and references.

**Table 4.28. The Theory of Post-Capitalism I.3: By Region— Ex: Spiritual/Communal in the Trans-Islamic Calling**

- **More Communal Than Individualistic**
  —*Collective*: building the webs of relationships to bind individuals
  —*Sharing*: cultivating the established "wisdom" through common experience
  —*Cooperative*: stressing harmony, solidarity, and commonality

- **More Informal-Legalistic Than Formal-Legalistic**
  —*Specific*: making efforts to know well the participants (family and larger community) in matters of common concern
  —*Affective*: mixing work with language and ritual on explicit religious (Islamic) ideals, texts, stories, and examples
  —*Ascriptive*: privileging local history and custom on relationships among kinship groups
  —*Particularistic*: preferring an unbiased insider with ongoing connections to all parties

- **More Spiritual Than Secular**
  —*Historical*: learning from the lessons of the past as a source of stability and guidance
  —*Deferential*: showing respect for age, experience, status, and leadership in communal affairs
  —*Honorable*: emphasizing face, dignity, prestige, and fairness
  —*Compassionate*: giving mercy and charity ("Zahah") to others

*Notes*: The categories and examples are solely illustrative (not exhaustive), and the comparison is relative (not absolute), nor are they mutually exclusive. As generalities, they allow exceptions. The specific forms of the trans-Islamic version need to be further developed in future after-postmodern history, as they will be different from the ones we now know, since the prefix "trans-" here means going beyond or deconstructing the Islamic values, while using them as the inspirational point at the beginning.
*Sources*: From Ch.10 of *FCD*. Refer to text for more info and references, especially from the works by George Irani (2000) and C. Murphy (September 19, 2001).

**Table 4.29. The Theory of Post-Capitalism I.4: By Universe— Ex: Spiritual/Communal in the Trans-Outerspace Calling**

- **More Communal Than Individual**
  —*Cooperative*: requiring teamwork in small space habitats
  —*Sharing*: learning from, and enjoying being with, each other in a small group in outer space

- **More Informal-Legalistic Than Formal-Legalistic**
  —*Specific*: knowing more about each other to facilitate living and working together in space, both as fellow astronauts and space-mates
  —*Affective*: being friendly and social to each other as vital to working and living in small space quarters
  —*Ascriptive*: nurturing comaraderie among fellow astronauts as if they are family members over time
  —*Particularistic*: building work relationship with enduring memory in a space mission

- **More Spiritual Than Secular**
  —*Long-Term*: looking beyond selfish materialistic concerns in a precarious space environment with potential life or death
  —*Loving/Caring*: cultivating deep bondage for the success of a long term space mission
  —*Transcendent*: searching for life meaning in outer space

*Notes*: The calling and examples in each category are solely illustrative (not exhaustive), since there will be many different outer-space value ideals in the distant future of space colonization. The comparison is also relative (not absolute), nor are they mutually exclusive. As generalities, they allow exceptions. And the specific forms of trans-outer-space calling need to be further developed in future after-postmodern history, as they will be different from the ones we now know, since the prefix "trans-" here means going beyond or deconstructing the current outer-space values, while using them as the inspirational point at the beginning. The point here is to solely give an extremely rough picture of a small part of the world to come that we still do not know much about.
*Source*: From Ch.10 of *FCD*. Refer to text for more info and references.

**Table 4.30. The Theory of Post-Capitalism II: Spiritual/Individualistic in the Post-Human Elitist Calling (Part I)**

- **More Individualistic Than Communal**
  —Setting up rank distinctions among unequals (e.g., between inferior humans and superior post-humans, or later among inferior post-humans and superior ones, relatively speaking)
  —Yearning for being not only distinguished from unequals, but also the first among equals (the best of the very best)
  —Recognizing the constraints of equality/inequality dialectics (or existential dialectics in general)

- **More Spiritual Than Secular**
  —Soul-searching for a high spiritual culture (not the trashy one for the masses). Mass culture is a dirty joke for them.
  —Exploring different spheres of non-human consciousness in the cosmos (something vastly superior than the human one)
  —Recognizing the constraints of freedom/unfreedom dialectics (or existential dialectics in general)

- **Qualifications**
  —Although post-human elitist post-democracy is comparable to post-human elitist post-capitalism in some respects, the former does not necessarily imply the latter (post-human elitist post-capitalism), just as democracy does not have to entail capitalism. They are two different (though related) entities.
  —But up to a certain threshold of incorporating government intervention with advanced info systems in future civilizations for higher spiritual concerns at the expense of the free market and materialist pursuit, the capitalist ideal will be overcome.
  —The overcome will not be Fascist or feudalistic, but post-capitalist, subject to the constraints of existential dialectics.

*(continued on next page)*

**Table 4.30. The Theory of Post-Capitalism II: Spiritual/Individualistic in the Post-Human Elitist Calling (Part II)**

*Notes*: The calling and examples in each category are solely illustrative (not exhaustive). The comparison is also relative (not absolute), nor are they mutually exclusive. As generalities, they allow exceptions. And the specific forms of post-human elitist post-capitalism need to be further developed in future after-postmodern history, as they will be different from the ones we now know, while using them as the inspirational point at the beginning. The point here is to solely give an extremely rough picture of a small part of the world to come that we still do not know much about.
*Sources*: From Ch.10 of *FCD* (and also *FPHC*, *BDPD*, and *BCPC*). Refer to the text for more info and references.

**Table 4.31. Capitalism, Non-Capitalism, and Post-Capitalism (Part I)**

---

• **Capitalism**

—*Theoretical Constructs*

•Allocation of scarce resources among alternative wants largely by free market for competition (whose characteristics in its ideal form include, for instance, no barrier to entry or exit, homogeneity, perfect information, a large number of buyers/sellers, and perfect factor mobility)

•More formal-legalistic than informal-legalistic, more individualistic than communal, and more material (secular) than spiritual

•Either (1) minimal government or (2) relatively active government

—*Types*

•Only (1): Different versions of market capitalism (e.g., the U.S.)

•Only (2): Different versions of welfare capitalism (e.g., Sweden)

• **Non-Capitalism**

—*Theoretical Constructs*

•Allocation of scarce resources among alternative wants mainly by political authority for policies (which can be regulative, redistributive, symbolic, and participatory)

•More informal-legalistic than formal-legalistic

•Either (1') more individualistic (for the elites), often (though not always) for material (secular) concerns, or (2') more communal (for the masses), often (though not always) for spiritual concerns

—*Types*

•Only (1'): Different versions on the Right (e.g., Fascist corporate-state economy for the glory of the new Rome, medieval lord-vassal-serf economy for the power of the feudalistic order)

•Only (2'): Different versions on the Left (e.g., Soviet command economy for the creation of the New Socialist Man)

---

*(continued on next page)*

**Table 4.31. Capitalism, Non-Capitalism, and Post-Capitalism (Part II)**

- **Post-Capitalism**
  —*Theoretical Constructs*
  - •Allocation of scarce resources among alternative wants largely by political authority with advanced info systems in future civilizations, subject to existential dialectics. In degree of allocating by authority, post-capitalism is more than capitalism but less than non-capitalism.
  - •More spiritual than secular (material)
  - •Either (1'') more individualistic or (2") more communal
  - •Like capitalism and non-capitalism, post-capitalism is also subject to the freedom/unfreedom and equality/inequality dialectics (or existential dialectics in general). There is no utopia, in the end; even were there one, dystopia would exist within it.
  - •Unlike capitalism and non-capitalism, post-capitalism makes use of a different degree of political authority with advanced info systems in future civilizations and strives for higher-spiritual cultures (especially in the post-human age), while acknowledging the constraints of existential dialectics and no longer valuing free market (as in capitalism) and economic control (as in non-capitalism) as sacred virtues.

  —*Types*
  - •Only (1"): Different versions of post-human elitist value ideals
  - •Only (2''): Different versions of trans-Sinitic value ideals
  - •Only (2''): Different versions of trans-feminine value ideals
  - •Only (2''): Different versions of trans-Islamic value ideals
  - •Only (2''): Different versions of trans-outerspace value ideals

*(continued on next page)*

**Table 4.31. Capitalism, Non-Capitalism, and Post-Capitalism (Part III)**

*Notes*: The calling and examples in each category are solely illustrative (not exhaustive). The comparison is also relative (not absolute), nor are they mutually exclusive. As generalities, they allow exceptions. And the specific forms of each calling need to be further developed in future after-postmodern history, as they will be different from the ones we now know, while using them as the inspirational point at the beginning. The point here is to solely give an extremely rough picture of a small part of the world to come that we still do not know much about.
*Source*: From Ch.10 of *FCD*. Refer to the text for more info and references.

**Table 4.32. Multiple Causes of the Emergence of Post-Capitalism (Part I)**

- **At the Micro-Physical Level**
  —Ex: intelligent life without the human physical-chemical system
  —Ex: mastering of quantum mechanics, electromagnetism, and other fields for the understanding of a broad range of anomalous experiences and the application for artificial intelligence for spiritual quest
  —Sources: Ch.7 of *FHC*; Chs.9-10 of *FCD*; Ch.1 of *FPHC*

- **At the Chemical Level**
  —Ex: space radiation and toxins
  —Sources: Ch.7 of *FHC*; Chs.9-10 of *FCD*

- **At the Bio-Psychological Level**
  —Ex: exo-biological evolution in deep space
  —Ex: genetic engineering of new beings
  —Ex: limits of human cognition
  —Ex: the rise of unfolding unconsciousness
  —Sources: Ch.2 & Chs.9-10 of *FCD*; Ch.7 of *FHC*; *FPHU*

- **At the Institutional Level**
  —Ex: the flawed logic of the free market
  —Ex: the need of a post-autistic economics
  —Sources: Ch.10 of *FCD*

- **At the Organizational Level**
  —Ex: the dark sides of formal-legalistic routines
  —Sources: Ch.3 of *FHC*; Ch.7 of *FCD*; Ch.3 of *FPHC*

- **At the Structural Level**
  —Ex: ever new forms of inequities, at home and abroad
  —Ex: the emergence of China, women, and Islam as major actors
  —Sources: Chs.5-6 of *FHC*; Chs.7, 9 & 10 of *FCD*; Chs.4-5 of *BDPD*

*(continued on next page)*

**Table 4.32. Multiple Causes of the Emergence of Post-Capitalism (Part II)**

---

- **At the Cultural Level**
  —Ex: freedom/unfreedom dialectics
  —Ex: equality/inequality dialectics
  —Sources: Ch.5 of *FHC*; Chs.3 & 10 of *FCD*; Ch.4 of *FPHC*; Ch.1 of *BDPD*

- **At the Systemic Level**
  —Ex: space habitats (in zero-gravity) and colonization
  —Ex: ultra advanced future info systems
  —Ex: qualitative demography
  —Sources: Ch.7 of *FHC*; Chs. 9 & 10 of *FCD*

- **At the Cosmological Level**
  —Ex: the colonization of multiverses
  —Ex: the alteration of space-time and the creation of new matter-energy
  —Ex: the expansion of floating consciousness
  —Ex: the spread of hyper-spatial consciousness
  —Sources: Ch.7 of *FHC*; Chs. 9 & 10 of *FCD*; Ch.4 of *FPHC*; *FPHST*

---

*Notes*: The examples in each category are solely illustrative (not exhaustive), and some of the items can be reclassified somewhere else. Nor are they always mutually exclusive. Since they are generalities, exceptions are expected.
*Sources*: Especially from *FHC*, *FCD*, *FPHC*, *BCPC*, *BDPD*, *FPHST*, and *FPHU*. See also other books and my perspectives on civilizational holism.

**Table 4.33. The Theoretical Debate on Civilization**

- **The Progressive Theory of Civilization**
  —*Thesis*: The "civilizing" process is "good," as opposed to the "barbarizing" process as something "bad," relatively speaking.
  —*Discourse*: Especially, though not exclusively, in the Enlightenment era and a bit before. Example: Thomas Hobbes—in that the tribes in primitive societies were "savages."

- **The Romantic Theory of Barbarity**
  —*Thesis*: The "civilizing" process is "bad," as opposed to the "barbarizing" process as something "good," relatively speaking.
  —*Discourse*: Especially, though not exclusively, in the Counter-Enlightenment circle. Example: Jean-Jacques Rousseau—in that civilization "corrupts" men, and the "savages" are in fact "noble."

- **The Moderate Theory of Civilization**
  —*Thesis*: The "civilizing" process is "good," but there is a price to pay, especially in systematic (compulsive) self-control.
  —*Discourse*: Especially, though not exclusively, in some late modern and postmodern circles. Example: Norbert Elias—in that social manners become more refined in the civilizing process, but self-control also becomes more systematic.

- **The Theory of Post-Civilization**
  —*Thesis*: The civilizing process is as evil and good as barbarity, and each cannot exist without the other, to be eventually superseded by post-civilization unto the post-human age.
  —*Discourse*: Proposed by Peter Baofu. See the rest of *BCIV* for more analysis.

*Source*: From *BCIV* on the theoretical debate

**Table 4.34. No Freedom Without Unfreedom in the Civilizing Processes (Part I)**

- **The Rationalizing Process (at the Level of Culture)**
  —if freer from the dominance of unreason (as in barbarism) in the civilizing process, then less free from the rationalizing process (be it in the form of the principle of either transcendence or immanence)
  —if freer from the principle of immanence in the rationalizing process, then less free from the inclination to commit terror in the name of reason and the relative underdevelopment of non-reason ( e.g., in relation to yoga and meditation)
  —if freer from the principle of transcendence in the rationalizing process, then less free from the relative underdevelopment of reason (e.g., in relation to systematic methodology) and the occurrence of oppression in the name of non-reason

- **The Pacifying Process (at the Level of Society)**
  —if freer from the dominance of pillage (as in savagery) in the civilizing process, then less free from the pacifying process (be it in the form of external control or self-control)
  —if freer from self-control in the pacifying process, then less free from the temptation of expansionist oppression and rebellious mindset in external control
  —if freer from external control in the pacifying process, then less free from the gruesome psychological self-torture and conformism in self-control

- **The Stewardizing Process (at the Level of Nature)**
  —if freer from the dominance of nature (as in the state of nature) in the civilizing process, then less free from the stewardizing process (be it in the form of the stewardship of creation or the covenant with nature)
  —if freer from the stewardship of creation in the stewardizing process, then less free from material underdevelopment, relatively speaking, and spiritual exclusion in the covenant with nature
  —if freer from the covenant with nature in the stewardizing process, then less free from ecological degradation and spiritual disconnection from nature in the stewardship of creation

*(continued on next page)*

**Table 4.34. No Freedom Without Unfreedom in the Civilizing Processes (Part II)**

---

- **The Subliming Process (at the Level of the Mind)**
  —if freer from the dominance of spontaneity (as in the wild state of the mind) in the civilizing process, then less free from the subliming process, be it in the form of (cyclical-centric) self-refinement or (linear-centric) self-discipline
  —if freer from (cyclical-centric) self-refinement in the subliming process, then less free from the (linear-centric) self-regimen (as a form of neurosis)
  —if freer from (linear-centric) self-discipline in the subliming process, then less free from the (cyclical-centric) self-torture (equally as a form of neurosis)

---

*Notes*: The examples in each category are solely illustrative (not exhaustive), and the comparison is relative (not absolute), nor are they necessarily mutually exclusive. And some can be easily re-classified elsewhere. As generalities, they allow exceptions.
Sources: From *BCIV*. See also *FHC*, *FCD*, *FPHC*, *BDPD*, and *BCPC*.

**Table 4.35. No Equality Without Inequality in the Civilizing Processes (Part I)**

- **The Rationalizing Process (at the Level of Culture)**
  —if more equal for the role of rationalization in the rationalizing process (of civilizational making), then less equal for that of mythicization (as in barbarism)
  —if more equal for the principle of transcendence in (linear-centric) rationalizing process, then less equal for the principle of immanence
  —if more equal for the principle of immanence in (cyclical-centric) rationalizing process, then less equal for the principle of transcendence

- **The Pacifying Process (at the Level of Society)**
  —if more equal for pacification in civilizational making, then less equal for the institution of pillaging and others (as in savagery)
  —if more equal for external control, relatively speaking, in pacifying process, then less equal for self-control
  —if more equal for self-control, relatively speaking, in pacifying process, then less equal for external-control

- **The Stewardizing Process (at the Level of Nature)**
  —if more equal for stewardship in the stewardizing process (of civilizational making), then less equal for reverent (submissive) existence (as in barbarism)
  —if more equal for the stewardship of creation in (linear- centric) stewardizing process, then less equal for the (cyclical-centric) covenant with nature for harmonious co-existence
  —if more equal for the (cyclical-centric) covenant with nature in the stewardizing process, then less equal for the (linear-centric) stewardship of nature for domination

*(continued on next page)*

**Table 4.35. No Equality Without Inequality in the Civilizing Processes (Part II)**

---

- **The Subliming Process (at the Level of the Mind)**
  —if more equal for the role of reason in the subliming process, then less equal for that of unreason (as in the natural state of wildness)
  —if more equal for the primacy of reason in (linear-centric) subliming process, then less equal for other faculties (e.g., intuition, existential feelings, and analogous thinking) in cyclical-centric one
  —if more equal for the exercise of other faculties (e.g., intuition, existential feelings, and analogous thinking) in cyclical-centric subliming process, then less equal for the role of reason in linear-centric counterpart

---

*Notes*: The examples in each category are solely illustrative (not exhaustive), and the comparison is relative (not absolute), nor are they mutually exclusive. And some can be easily reclassified else-where. As generalities, they allow exceptions.
*Sources*: From *BCIV*. See also *FHC*, *FCD*, *FPHC*, *BDPD*, and *BCPC*.

**Table 4.36. Five Theses on Post-Civilization**

• Post-civilization no longer treats civilization as good and barbarity as evil (relatively speaking), nor does it nostalgically regard barbarity as good and civilization as evil (relatively speaking again). Civilization is as evil and good as barbarity.

• Post-civilization also no longer accepts the dichotomy between civilization and barbarity. Civilization cannot exist without barbarity. It is no longer necessary to preserve civilization, any more than it is imperative to destroy barbarity. To go beyond civilization and barbarity is to acknowledge the co-existence of both, although the degree of scaling one over the over varies from case to case—but is not to be extreme in largely favoring one over the other, *on average* (subject to the constraints of existential dialectics).

• Post-civilization is thus subject to the constraints of existential dialectics. There is no freedom without unfreedom, and no equality without inequality, for instance. There will be no utopia; even should there be one, there would be dystopia embedded within it.

• Post-civilization will eventually replace civilization (as a form of life settlement), to be dominated by post-capitalist and post-democratic lifeforms here on earth and in deep space (besides other alien lifeforms that we have never known), unto the post-human age in multiverses. Those few post-humans who keep civilization will live in a "post-human civilization," while the rest (the majority), who choose post-civilization, will evolve towards the state of "post-human post-civilization." One therefore should not confuse "post-human civilization" with "post-human post-civilization," as the two are not the same.

• Post-civilization will confront psychosis as a primary problem in the culture of virtuality unto the post-human age, just as civilization has neurosis as a primary one of its own (although both neurosis and psychosis are major problems in both).

*Notes*: The comparison in each category is relative (not absolute), nor are they necessarily mutually exclusive. And some can be easily re-classified elsewhere. As generalities, they allow exceptions.
*Sources*: From *BCIV*. See also *FHC*, *FCD*, *FPHC*, *BDPD*, and *BCPC*.

**Table 4.37. Barbarity, Civilization, and Post-Civilization**

- **The Rationalizing Process (at the Level of Culture)**
  —*Barbarity*
    - More mythicizing than rationalizing, relatively speaking
  
  —*Civilization*
    - More rationalizing than mythicizing, relatively speaking
  
  —*Post-Civilization*
    - Beyond the dichotomy, subject to existential dialectics

- **The Pacifying Process (at the Level of Society)**
  —*Barbarity*
    - More pillaging than pacifying, relatively speaking
  
  —*Civilization*
    - More pacifying than pillaging, relatively speaking
  
  —*Post-Civilization*
    - Beyond the dichotomy, subject to existential dialectics

- **The Stewardizing Process (at the Level of Nature)**
  —*Barbarity*
    - More revering than stewardizing, relatively speaking
  
  —*Civilization*
    - More stewardizing than revering, relatively speaking
  
  —*Post-Civilization*
    - Beyond the dichotomy, subject to existential dialectics

- **The Subliming Process (at the Level of the Mind)**
  —*Barbarity*
    - More impulsing than subliming, relatively speaking
  
  —*Civilization*
    - More subliming than impulsing, relatively speaking
  
  —*Post-Civilization*
    - Beyond the dichotomy, subject to existential dialectics

*Notes*: The comparison in each category is relative (not absolute), nor are they necessarily mutually exclusive. And some can be easily re-classified elsewhere. As generalities, they allow exceptions.
*Sources*: From *BCIV*. See also *FHC*, *FCD*, *FPHC*, *BDPD*, and *BCPC*.

**Table 4.38. Types of Super Civilization in the Cosmos (Part I)**

- **Type I**
  —a civilization which gains control of and uses the total energy output "falling on its planet from its sun for interstellar communication" (or, in general, space colonization). For N. Kardashev, who proposed the first three types, human civilization is currently Type Zero (Type O), which is below even Type I, since its present energy consumption for all purposes, let alone for interstellar communication, is still 10,000 times less.

- **Type II**
  —a civilization which gains control of and uses directly the total energy output of its sun for interstellar communication (or, in general, space colonization).

- **Type III**
  —a civilization which gains control of and uses the total energy output of its galaxy for interstellar communication (or, in general, space colonization).

- **Type IV**
  —a civilization which gains control of and uses the total energy output of its cluster of galaxies for interstellar communication (or, in general, space colonization).

- **Type V**
  —a civilization which gains control of and uses the total energy output of its supercluster of galaxies for interstellar communica-tion (or, in general, space colonization).

*(continued on next page)*

**Table 4.38. Types of Super Civilization in the Cosmos (Part II)**

- **Type…n**
  —So continues the series in what I call the cyclical progression of hegemony in the cosmos and beyond.

*Notes*: The Russian astrophysicist Nikolai Kardashev proposed the first three types of super civilization in terms of total energy out-put for interstellar communication. (CSM 1979) I extend his argument further to propose Type IV, Type V, Type VI, and Type…n, in the context of my claim about the cyclical progression of he-gemony in the cosmos and beyond.
*Sources*: From *Table 9.4* of *FCD*. See *FHC*, *FCD*, and *FPHC* for more info.

**Table 4.39. The Civilizational Project from Pre-Modernity to After-Postmodernity (Part I)**

| | *Pre-Modern* | *Modern* | *Postmodern* | *After-Postmodern* |
|---|---|---|---|---|
| *Main narratives* | •Sacralness<br>•Courtliness<br>•Vitalism<br>•Animism | •Freedom<br>•Equality<br>•Fraternity | •Multiplicity<br>•Hybridization | •Naked contingency<br>•Cyclical progres-sion of hegemony |
| *Main institutions* | •Monarchy<br>•Aristocracy<br>•Feudalism<br>•Holy order<br>•Primitivism | •Capitalism<br>•Liberalism<br>•Socialism<br>•Nazism<br>•Fascism | •Capitalism<br>•Liberalism<br>•Postmodern politics of difference | •Post-Capitalism<br>•Post-Democracy<br>•Others |
| *Main techno-logical and economic revolutions* | •Agricultural | •Service<br>•Industrial | •Informational | •Biological<br>•Material<br>•Energy<br>•Space<br>•Others |

*(continued on next page)*

**Table 4.39. The Civilizational Project from Pre-Modernity to After-Postmodernity (Part II)**

| | *Pre-Modern* | *Modern* | *Postmodern* | *After-Postmodern* |
|---|---|---|---|---|
| *Main agents* | •Males<br>•Upper strata<br>•Mini-states | •Males<br>•Upper strata<br>•Whites<br>•Empires | •Males<br>•Upper strata<br>•Whites<br>•Others<br>•Supra-states<br>•IO's | •Post-humans<br>•Humans<br>•Others |
| *Main impacts* | •Local | •Inter-national | •Global | •Outer-space<br>•Multiverse |
| *Main outcomes* | •Towards moderntiy<br>•Rise of linear- & cyclical-centric civiliza-tions | •Towards post-modernity<br>•Dominance of linear-centric civilization | •Towards after-post-moderntiy<br>•Linear-centric civiliza-tion in crisis | •Towards human (& maybe post-human) extinction<br>•Rise of post-civiliza-tion, especially in post-human forms of space-time |

*Notes*: The examples in each category are solely illustrative (not exhaustive) nor necessarily mutually exclusive, and the comparison is relative (not absolute). As generalities, they allow exceptions.
*Sources*: From *Table 10.16* of *FCD*—and also from *BCIV* on post-civilization (and *FPHST*)

**Table 4.40. Civilizational Holism (Part I)**

---

• **At the Micro-Physical Theoretical Level**
—Ex: Mastering of quantum mechanics, electromagnetism, and other fields for the understanding of a broad range of anomalous experiences and theapplication for artificial intelligence (*Sec.1.4.1* of *FPHC*)

• **At the Chemical Theoretical Level**
—Ex: Unprecedented expansion of (and violence to) the mind through ever new forms of drugs (and virtual technologies, for that matter) (Ch.9 of *FCD*)

• **At the Biological Theoretical Level**
—Ex: Humans are not biologically equal, on the basis of race, gender, ethnicity, age, and whatnot. (*Sec.2.6* & Ch.10 of *FCD*; *BNN*) And post-humans will experience the same fate, in an even more amazing way.

• **At the Psychological Theoretical Level**
—Ex: Human cognitive impartiality and emotional neutrality are quite limited. (*Secs.2.4-2.5* of *FCD*)
—Ex: Rise of Floating Consciousness (Ch.10 of *FCD*; Chs.1 & 4 of *FPHC*) and Unfolding Unconsciousness (*FPHU*)

• **At the Organizational Theoretical Level**
—Ex: Administrative colonization of deep space, with less legal-formalism in some corners. (Chs.9-10 of *FCD*)

• **At the Institutional Theoretical Level**
—Ex: Both capitalism and democracy will not last, to be superseded by different versions of post-capitalism and post-democracy in after-postmodernity. (Ch.10 of *FCD*)

---

*(continued on next page)*

**Table 4.40. Civilizational Holism**
**(Part II)**

- **At the Structural Theoretical Level**
  —Ex: Social stratification reappears in ever new forms, also with new causes and new players in the cyclical progression of hegemony. (Chs.8-10 of *FCD*)
  —Ex: The world of hyper-empires, and the union of the unions (*BWT*)

- **At the Systemic Theoretical Level**
  —Ex: Outerspace expansion: local → regional → global → solar → galactic → clustery → multiversal (Ch.9 of *FCD*)
  —Ex: Demographic transition: human extinction, and the rise of post-humans (e.g., cyborgs, thinking machines, thinking robots, genetically altered superior beings, floating consciousness, hyper-spatial consciousness) (Ch.4 of *FPHC*; Ch.10 of *FCD*; & Ch.7 of *FHC*)
  —Ex: New technological forces in material sciences, electronic and communication sciences, energy sciences, biosciences, manufacturing and engineering sciences, and space sciences (Ch.10 of *FCD* & Ch.7 of *FHC*)
  —Ex: Systematic dominance towards nature for space colonization (Chs.9-10 of *FCD*; Chs.2 & 7 of *FHC*)

- **At the Cultural Theoretical Level**
  —Ex: The post-human transcendence of freedom and equality (Ch.10 of *FCD*)
  —Ex: Methodological Holism (Ch.1 of *FCD*; Ch.1 of *FPHC*; *Sec.2.1* & *Sec.2.5* of *BCPC*)
  —Ex: The Evolution from Barbarity to Post-Civilization (*BCIV*)

- **At the Cosmological Theoretical Level**
  —Ex: Mastering of dark matter and dark energy, and the exploration of multiverses (Ch.4 of *FPHC*; Ch.10 of *FCD*; & Ch.7 of *FHC*)
  —Ex: Alternation of space-time (*FPHST*)
  —Ex: The emergence of hyper-spatial consciousness (*FPHC*)

*(continued on next page)*

**Table 4.40. Civilizational Holism (Part III)**

- **At Other Levels**
  —Ex: Historical: pre-modernity → modernity → postmodernity → after-postmodernity (human distinction, and the rise of post-humans, including floating consciousness, hyper-spatial consciousness, and unfolding unconsciousness) (Ch.7 of *FHC*; Ch.10 of *FCD*; *FPHC*; *FPHU*)

*Notes*: These examples are solely illustrative (not exhaustive), and some of the items can be reclassified somewhere else. Nor are they always mutually exclusive. Since they are generalities, exceptions are expected. And the comparison is relative, not absolute.
*Sources*: From *Table 5.1* of *FPHC*—with details from *FHC*, *FCD*, and the rest of my books. Not every aspect in each category as discussed in all my books are presented here, since there are too many issues. For more info, also consult the table on theories on civilizational holism and, of course, the books themselves.

**Table 4.41. Theories on Civilizational Holism (Part I)**

---

**I. Theories in Relation to Nature**

—*At the Macro-Physical (Cosmological) Theoretical Level*

• 34. Theory of Hyper-Spatial Consciousness (Peter Baofu) (Ch.4 of *FPHC*; *FPHG*)
(• 32). Selective Theory of Geometry (Peter Baofu) (*FPHG*)
(• 31). Perspectival Theory of Space-Time (Peter Baofu) (*FPHST*)
(• 30). Dialectic Theory of Complexity (Peter Baofu) (*FC*)
(• 8). Theory of Floating Consciousness (Peter Baofu) (Ch.10 of *FCD*; Chs.1 & 4 of *FPHC*)
• 33. Theory of the Geopower of Nature (Peter Baofu) (Ch.4 of *ALD*)

—*At the Micro-Physical Theoretical Leve*

• 32. Selective Theory of Geometry (Peter Baofu) (*FPHG*)
• 31. Perspectival Theory of Space-Time (Peter Baofu) (*FPHST*)
• 30. Dialectic Theory of Complexity (Peter Baofu) (*FC*)

**II. Theories in Relation to Culture**

—*At the Cultural Theoretical Level*

• 29. Comparative Theory of Religion (Peter Baofu) (Ch.3 of *FHC*; Ch.9 of *FCD*; Ch.1 of *FPHK*)
• 28. Theory of Post-Civilization (Peter Baofu) (*BCIV*)
• 27. Theory of the Trinity of Modernity to Its After-Postmodern Counterpart (Peter Baofu) (*FHC*; Ch.10 of *FCD*)
• 26. Transformative Theory of Aesthetic Experience (Peter Baofu) (*FAE*)

---

*(continued on next page)*

**Table 4.41. Theories on Civilizational Holism (Part II)**

**III. Theories in Relation to Society**

—*At the Organizational Theoretical Level*

- 25. Theory of E-Civic Alienation (Peter Baofu) (Ch.7 of *FCD*)
- 24. Combinational Theory of Organizational Behavior (Peter Baofu) (Ch.6 of *ALD*)

—*At the Institutional Theoretical Level*

- 23. Theory of Post-Capitalism (Peter Baofu) (Ch.10 of *FCD*; Chs.2 & 4 of *FPHC*; *BCPC*)
- 22. Theory of Post-Democracy (Peter Baofu) (Ch.10 of *FCD*; Chs.3 & 4 of *FPHC*; *BDPD*)
- 21. Dynamic Theory of Comparative Political Systems (Peter Baofu) (*ALD*)

—*At the Systemic Theoretical Level*

- 20. Contingent Theory of Urban Planning (Peter Baofu) (*FPHUP*)
- 19. Totalistic Theory of Communication (Peter Baofu) (*FPHMM*; *FCD*; *FHC*)
- 18. Ambivalent Theory of Technology (Peter Baofu) (*FPHE*; *FCD*; *FHC*)
- 17. Multifaceted Theory of War and Peace (Peter Baofu) (Ch.9 of *FCD*; Ch.1 of *FPHK*)
- 16. Theory of Post-Humanity (Peter Baofu) (Ch.7 of *FHC*; Chs.3, & 10 of *FCD*; Chs.1, 3 & 4 of *FPHC*; and other books of mine)
- 15. Theory of the Cyclical Progression of System Integration and Fragmentation (Peter Baofu) (Chs.9-10 of *FCD*)
- 14. Synthetic Theory of Information Architecture (Peter Baofu) (*FIA*)

*(continued on next page)*

**Table 4.41. Theories on Civilizational Holism (Part III)**

---

**III. Theories in Relation to Society (*cont'd*)**

—*At the Structural Theoretical Level*

- 13. Theory of the Cyclical Progression of Hegemony (Peter Baofu)
  (Chs.9-10 of *FCD*; Chs.1, 3 & 4 of *FPHC*; *BDPD*)
- 12. Theory of the Cyclical Progression of Empire-Building (Peter Baofu)
  (*BWT*)

**IV. Theories in Relation to the Mind**

—*At the Biological Theoretical Level*

- 11. Theory of Contrastive Advantages (Peter Baofu)
  (*Sec.2.6* & Ch.10 of *FCD; BNN*)

—*At the Psychological Theoretical Level*

- 10. Comprehensive Theory of Creative Thinking (Peter Baofu)
  (*FPHCT*)
- 9. Theory of Unfolding Unconsciousness—also known as the Unfolding Theory of Anomalous Experience (Peter Baofu)
  (*FPHU*)
- 8. Theory of Floating Consciousness (Peter Baofu)
  (Ch.10 of *FCD*; Chs.1 & 4 of *FPHC*)
- 7. Theory of Cognitive Partiality (Peter Baofu)
  (*Sec.2.4* of *FCD*; *Sec.4.5.1.1* of *BCPC*)
- 6. Theory of Emotional Non-Neutrality (Peter Baofu)
  (*Sec.2.5* of *FCD*; *Sec.4.5.2* of of *BCPC*)
- 5. Theory of Behavioral Alteration (Peter Baofu)
  (*Sec.4.5.3* of *BCPC*)

---

*(continued on next page)*

**Table 4.41. Theories on Civilizational Holism (Part IV)**

**V. Meta-Theories (in Relation to Theories)**
—*At the Ontological Meta-Theoretical Level*
- 4. Theory of Existential Dialectics, or the Holistic Theory of Knowledge (Peter Baofu) (*FHC*; *FCD*; *FPHC*; *BDPD*; *FC*; *FAE*; *ALD; FIA; FPHK*; etc,)
- 3. Contrastive Theory of Rationality (Peter Baofu) (*FPHML*)

—*At the Methodological Meta-Theoretical Level*
- 2. Theory of Methodological Holism (Peter Baofu) (Ch.1 of *FCD*; Ch.1of *FPHC*; *Sec.2.1* & *Sec.2.5* of *BCPC*; *FC*; *FPHK*; etc.)

**VI. Theories in Relation to the Rest**
—*At Other Levels (Historical)*
- 1. Theory of the Evolution from Pre-Modernity to After-Postmodernity (Peter Baofu) (*FHC*; Ch.9-10 of *FCD*; *FPHC*)

*Notes*: All these theories are my constructions, as some of the main contributions of my grant project on civilization and its future. These examples are solely illustrative (not exhaustive), and some of the items can be reclassified somewhere else. Nor are they always mutually exclusive. Since they are generalities, exceptions are expected.
*Sources*: From *FHC*, *FCD*, *FPHC*, *BDPD*, *BCPC*, *BCIV*, *FPHST*, *BNN*, *BWT*, *FC*, *FAE*, *ALD*, *FIA*, *FPHU*, *FPHK*, *FPHML*, *FPHE*, *FPHMM*, *FPHCT*, *FPHG* & *FPHUP*

# BIBLIOGRAPHY


*AAAS*. 2009. "Forest Products" (January 05). <http://atlas.aaas.org/pdf/63-66.pdf>.

Alexander, R.McNeill. 2001. "Design by Numbers." *Nature*, v.412 (August 9).

Allen, Julia & Douglas F. Barnes. 1985. "The Causes of Deforestation in Developing Countries." *Annals of the Association of American Geographers*, 75: 2163-184.

Anderson, L.; L. Goodman; B. Mulligan; & H. Edward & Regen. 1983. "Effects of Sounds on Preferances for Outdoor Settings." *Environment and Behavior*, vol. 15, no. 5: 539-566.

Angelsen, Arild; & David Kaimowitz. 1999. "Rethinking the Causes of Deforestation: Lessons from Economic Models." *The World Bank Research Observer*.14:1.

Baofu, Peter. 2009. *The Future of Post-Human Geometry: A Preface to a New Theory of Infinity, Symmetry, and Dimensionality*. Cambridge, England: Cambridge Scholars Publishing.

______.2009a. *The Future of Post-Human Creative Thinking: A Preface to a New Theory of Invention and Innovation*. Cambridge, England: Cambridge Scholars Publishing.

______.2009b. *The Future of Post-Human Mass Media: A Preface to a New Theory of Communication*. Cambridge, England: Cambridge Scholars Publishing.

______.2009c. *The Future of Post-Human Engineering: A Preface to a New Theory of Technology*. Cambridge, England: Cambridge Scholars Publishing.

______.2008. *The Future of Post-Human Mathematical Logic: A Preface to a New Theory of Rationality*. Cambridge, England: Cambridge Scholars Publishing.

______.2008a. *The Future of Post-Human Knowledge: A Preface to a New Theory of Methodology and Ontology*. Oxford, United Kingdom: Chandos Publishing (Oxford) Ltd.

______.2008b. *The Future of Post-Human Unconsciousness: A Preface to a New Theory of Anomalous Experience*. Cambridge, England: Cambridge Scholars Publishing.

______.2008c. *The Future of Information Architecture: Conceiving a New Way to Understand Taxonomy, Network, and Intelligence*. Oxford, United Kingdom: Chandos Publishing (Oxford) Ltd.

_____.2007. *The Rise of Authoritarian Liberal Democracy: A Preface to a New Theory of Comparative Political Systems*. Cambridge, England: Cambridge Scholars Publishing.

_____.2007a. *The Future of Aesthetic Experience: Conceiving a Better Way to Beauty, Ugliness, and the Rest*. Cambridge, England: Cambridge Scholars Publishing.

_____.2007b. *The Future of Complexity: Conceiving a Better Way to Understand Order and Chaos*. London, United Kingdom: World Scientific Publishing Co.

_____.2007c. *Beyond the World of Titans, and the Remaking of World Order: A Preface to a New Logic of Empire-Building*. Cambridge, England: Cambridge Scholars Publishing.

_____.2006. *Beyond Nature and Nurture: Conceiving a Better Way to Understand Genes and Memes*. Cambridge, England: Cambridge Scholars Publishing.

_____.2006a. *The Future of Post-Human Space-Time: Conceiving a Better Way to Understand Space and Time*. New York: Peter Lang Publishing, Inc.

_____.2006b. *Beyond Civilization to Post-Civilization: Conceiving a Better Model of Life Settlement to Supersede Civilization*. New York: Peter Lang Publishing, Inc.

_____.2005. *Beyond Capitalism to Post-Capitalism: Conceiving a Better Model of Wealth Acquisition to Supersede Capitalism*. New York: The Edwin Mellen Press.

_____.2004. *The Future of Post-Human Consciousness*. New York: The Edwin Mellen Press.

_____.2004a. *Beyond Democracy to Post-Democracy: Conceiving a Better Model of Governance to Supersede Democracy* (2 volumes). New York: The Edwin Mellen Press.

_____.2002. *The Future of Capitalism and Democracy*. Maryland: The University Press of America.

_____.2000. *The Future of Human Civilization* (2 volumes). New York: The Edwin Mellen Press.

Barrow, John D. 1998. *Impossibility: The Limits of Science and the Science of Limits*. Oxford: Oxford University Press.

*BBC.com* (BBC). 2009. "The Tropical Rainforest" (January 05). <http://www.bbc.co.uk/schools/gcsebitesize/geography/ecosystems/ecosystemsrainforestrev1.shtml>.

_____.2009a. "Rainforest Water and Nutrient Cycles" (January 05). <http://www.bbc.co.uk/schools/gcsebitesize/geography/ecosystems/ecosystemsrainforestrev4.shtml>.

Bellah, Robert. 1985. "The Quest for the Self: Individualism, Morality, Politics"

(January 23-26). Claremont, CA: Claremont Institute Conference. <http://campus.udayton.edu/~amu-usa/pdfs/bellah_1.pdf>.

Bellamy, Edward. 1888. *Looking backward, 2000-1887*. Boston: Ticknor and Company.

Benhabib, S., ed. 1996. *Democracy and Difference: Contesting the Boundaries of the Political*. Princeton, NJ: Princeton University Press.

Benjamin, W. 1986. "A Berlin Chronicle." *Reflections: Essays, Aphorisms, Autobiographical Writings*. P. Demetz, ed. New York: Schocken.

Benson, John F. & Maggie Roe. 2000. "The Scale and Scope of Landscape and Sustainability." *Landscape and Sustainability*. John F. Benson and Maggie H. Roe. eds. London, UK: Spon.

Berry, David. 1976. "Preservation of Open Space and the Concept of Value." *American Journal of Economics and Sociology*: 113-124.

Betsky, A. 1997. *Queer Space: Architecture and Same-Sex Desire*. New York: William Morrow & Co.

Biletzki, Anat; & Anat Matar. 2006. "Ludwig Wittgenstein" (Nov. 17). *Stanford Encyclopedia of Philosophy*. <http://plato.stanford.edu/entries/wittgenstein/>.

*Bio-Medicine* (BM) 2007. "Single-Largest Biodiversity Survey Says Primary Rainforest Is Irreplaceable" (November 14). <http://www.bio-medicine.org/biology-news-1/Single-largest-biodiversity-survey-says-primary-rainforest-is-irreplaceable-1218-1/>.

Black, Richard. 2008. "Nature Loss 'to Hurt Global Poor'" (May 29). *BBC News*. <http://news.bbc.co.uk/2/hi/science/nature/7424535.stm>.

Borch-Jacobsen, Mikkel. 2000. "How Fabrications Differ from a Lie" (April 13). *London Review of Books*. <http://www.lrb.co.uk/v22/n08/borc01_.html>.

Brydson, Nicole.2008. "Brooklyn, The Borough: A Case of Gentrification" (May 23). *The New York Observer*. <http://www.observer.com/2008/brooklyn-borough-16>.

Bunge, Mario. 1999. *Dictionary of philosophy*. Amherst, MA: Prometheus Books.

*Bureau of Municipal Research* (BMR). 1971. "Urban Open Space: Luxury or Necessity?"

Butler, Rhett. 2007. "Forest Disturbance Reduces Biodiversity in the Amazon Rainforest" (July 02). *Mongabay.com*. <http://news.mongabay.com/2007/0702-gardner.html>.

______.2007a. "Destruction of Renewable Resources." *Mongabay.com*. <http://rainforests.mongabay.com/0905.htm>.

Callon, M. 1999. "The Role of Lay People in the Production and Dissemination of Scientific Knowledge. Science." *Technology & Society*, 4: 81-94.

Campbell, D. 1975. "On the Conflicts between Biological and Social Evolution and between Psychology and Moral Tradition," *American Psychologist*.

Washington, D.C.: The American Psychological Association, Inc. (December), vol.30, n.12.

Caro, Robert A. 1974. *The Power Broker: Robert Moses and the Fall of New York*. New York: Vintage Books.

Chase, Scott. 1993. "Tachyons" (March). *Physics FAQ*. <http://math.ucr.edu/home/baez/physics/ParticleAndNuclear/tachyons.html>.

Coase, Ronald. 1988. *The Firm, the Market and the Law*. Chicago, IL: University of Chicago Press.

Coles, Robert. 1980. "Civility and Psychology" (Summer). *Daedalus*: 136-137.

Corazzon, Raul. 2007. "Definitions of Ontology: From Nicolai Hartmann to the Present Time." <http://www.formalontology.it/section_4_two.htm>.

Corbett, Dan. 2008. "Introduction to String Theory" (November 23). *ThinkQuest*. <http://library.thinkquest.org/27930/stringtheory1.htm>.

Corburn, Jason. 2008. The session on "Sustaining Cities: Environment, Economic Development, and Empowerment" (April 04). *M.I.T. World*. <http://mitworld.mit.edu/video/574>.

Costanza, R. & C. Folke. 1996. "The Structure and Function of Ecological Systems in Relation to Property-Rights Regimes." *Rights to Nature*. S.S. Hanna, C. Folke, and K-G. Maler, eds. Washington, DC: Island Press.

_____.& B.C. Patten. 1995. "Defining and Predicting Sustainability." *Ecological Economics*, 15: 193-196.

_____.1992. *Toward An Operational Definition of Ecosystem Health in Ecosystem Health*. R. Costanza, B. G. Norton, and B. D. Haskell, eds. Washington, DC: Island Press.

Cunningham, Chris & Martin Auster. 1992 *Urban Design and Regulation*. Armidale: Department of Geography and Planning, University of New England.

D'Agostino, Bill. 2005. "Foothills Park Will Remain Closed" (October 25). *Palo Alto Online News*. <http://www.paloaltoonline.com/news/show_story.php?id=1941>.

Damasio, Antonio. 1994. *Descartes' Error: Emotion, Reason, and the Human Brain*. New York: Putnam Publishing.

Dempster, M. Beth. 1998. "A Self-Organizing Systems Perspective on Planning For Sustainability." Waterloo, Ontario, Canada. <http://www.nesh.ca/jameskay/ersserver.uwaterloo.ca/jjkay/grad/bdempster/>.

Denman, Kenneth; G. Brasseur; A. Chidthaisong; P. Ciais; P.M. Cox; R.E. Dickinson; D. Hauglustaine; C. Heinze; E. Holland; D. Jacob; U. Lohmann; S. Ramachandran; P.L. da Silva Dias; S.C. Wofsy & X. Zhang. 2007. "Couplings Between Changes in the Climate System and Biogeochemistry." *Climate Change 2007: The Physical Science Basis. Contribution of Working*

*Group I to the Fourth Assessment Report of the Intergovernmental Panel on Climate Change*. S. Solomon, D. Qin, M. Manning, Z. Chen, M. Marquis, K.B. Averyt, M.Tignor, and H.L. Miller, eds. Cambridge, UK: Cambridge University Press. <http://www.ipcc.ch/pdf/assessment-report/ar4/wg1/ar4-wg1-chapter7.pdf>.

Dhammika, S. 2009. "THE BROKEN BUDDHA: Critical Reflections on Theravada and a Plea for a New Buddhism." <http://www.buddhistische-gesellschaft-berlin.de/downloads/brokenbuddhanew.pdf>.

Dixon, J. A. & L. A. Fallon. 1989. "The Concept of Sustainability: Origins, Extensions and Usefulness for Policy." *Society and Natural Resources*, 2: 73-84.

Dokoupil, Tony. 2009. "America's Top Killer: Us" (January 02). *Newsweek*. <http://www.newsweek.com/id/177587/page/1>.

Domhoff, William. 1983. "Urban Renewal: How Corruption Operates Locally." *Who Rules America Now?* New York: Prentice-Hall, Inc. <http://www.worldfreeinternet.net/news/urban.htm>.

Doubleday, Nancy C. 1993. "Finding Common Ground: Natural Law and Collective Wisdom." *Traditional Ecological Knowledge*. J. T. Inglis, ed. Australia: International Development Resource Centre.

Dovers, S. R. & J. W. Handmer. 1992. "Uncertainty, Sustainability and Change." *Global Environmental Change*, 2(1): 262-276.

Downs, Anthony. 1957. *An Economic Theory of Democracy*. New York, Harper.

Emerson, Ralph Waldo. 1983. *Essays and Lectures*. N.Y.: The Library of America.

Engwicht, David. 2009. "The Connection between Religion and Urban Planning" (January 01). Research paper written as part of studies at Griffith University. Brisbane, Australia. <http://www.lesstraffic.com/Articles/Place/Religion.pdf>.

Everett, Hugh. 1957. "'Relative State' Formulation of Quantum Mechanics" (July). *Reviews of Modern Physics*, vol. 29, no.3. <http://www.univer.omsk.su/omsk/Sci/Everett/paper1957.html>.

Eysenbach, Mary. 2008. "Park System Function and Services." *From Recreation to Re-creation*. American Planning Association.

Eysenck, Hans. 1986. *Decline and Fall of the Freudian Empire*. New York: Viking.

Fearnside1, Philip; & William F. Laurance. 2004. "Tropical Deforestation and Greenhouse-Gas Emissions" (August). *Ecological Applications*, v.14, #4: 982–986.

Firehammer, Reginald. 2005. "Basic Principles of Ontology" (January 05). <http://usabig.com/autonomist/philosophy/ontology2.html>.

Fishman, Robert. 1977. *Urban Utopias in the Twentieth Century: Ebenezer*

*Howard, Frank Lloyd Wright, and Le Corbusier*. Cambridge, England: The MIT Press.

*Fondation Chirac* (FC). 2008. “Deforestation and Desertification.” <http://www.fondationchirac.eu/en/deforestation/>.

Fitzpatrick, Dan. 2000. “The Story of Urban Renewal: In East Liberty and Elsewhere, Pittsburgh's Dominant Public Policy Tool Didn't Work Out as Planned” (Sunday, May 21). Post-Gazette.com. <http://www.post-gazette.com/businessnews/20000521eastliberty1.asp>.

Frenchman, Dennis; Antonio di Mambro & Martha Welborne. 2008. “Building Responsive Cities: Technology, Design, and Development” (April 04). *M.I.T. World*. <http://mitworld.mit.edu/video/580>.

Friedmann, John. 1996. *Two Centuries of Planning Theory: An Overview in Explorations in Planning Theory*. S. J. Mandlebaum, L. Mazza, and R. W. Burchell, eds. Center for Urban Policy Research.

_____.1993. “Toward a Non-Euclidean Mode of Planning.” *American Planning Association Journal*, 59(4): 482-486.

_____.1987. *Planning in the Public Domain: From Knowledge to Action*. NJ: Princeton University Press.

Forester, John. 1989. *Planning in the Face of Power*. CA: University of California Press.

Fromm, E. 1975. *Man for Himself*. New York: Fawcett Premier.

Frost, Kathryn; & Frank Richardson. 2009. “Hate, Individualism, and the Social Bond.” *International Network on Personal Meaning*. <http://www.meaning.ca/archives/archive/pdfs/frost-hate.pdf>.

Fruth, Barbara. 2009. “Medicine from the Rainforest” (January 05). <http://www.bmbf.de/en/12484.php>.

Gadgil, M. & F. Berkes. 1991. “Traditional Resource Management Systems *Resource Management and Optimization*, 8(3-4): 127-141.

Gebbie, Kristine. 2008. “Urban Health.” *Encyclopedia of Public Health*. <http://www.answers.com/topic/urban-health>.

Geist, Helmut & Eric F. Lambin. 2002. “Proximate Causes and Underlying Driving Forces of Tropical Deforestation.” *BioScience*, 52:2 143-150.

Gethin, Rupert, tr. 2008. *Sayings of the Buddha: A Selection of Suttas from the Pali Nikāyas*. Oxford, UK: Oxford University Press.

Giedion, Sigfried. 1941. *Space, Time and Architecture; the Growth of a New Tradition*. Cambridge, MA: The Harvard University Press.

Gill, Don & Penelope Bonnett. 1973. *Nature in the Urban Landscape: A Study of City Ecosystems*. Baltimore, MD: York Press.

Goering, Paula & Tammy Morrell-Bellai. 1998. “Pathways Project: Integrating Qualitative and Quantitative Findings in Describing Pathways to

Homelessness." *Proceedings and Recommendations on "Mental Illness and Pathways into Homelessness: Findings and Implications"* (January 16). Mental Health Policy Research Group. <http://www.camh.net/hsrcu/html_documents/pathways_proceedings.html>.

Goldberg, M A. 1985. "The Irrationality of 'Rational' Planning: Exploring Broader Bases for Planning and Public Decisionmaking." *Rationality in Planning: Critical Essays on the Role of Rationality in Urban and Regional Planning.* M. Breheny and A. Hooper, eds. London: Pion.

*Google.com* (GL). 2008. "Brasilia" (December 15) <http://maps.google.com/?ie=UTF8&om=1&z=12&ll=-15.79787,-47.896271&spn=0.143704,0.233459&t=k>.

Gott, J. Richard. 2002. *Time Travel in Einstein's Universe*. Boston : Houghton Mifflin.

Grant, Benjamin. 2003. "What is Gentrification?" (June 17). *Public Broadcasting Service*. <http: //www.pbs.org/pov/pov2003/flagwars/special_gentrification.html>.

Gratz, Roberta Brandes. 1989. *The Living City*. New York: Simon and Schuster.

Green, David & David Newth. 2001. "Towards a Theory of Everything? – Grand Challenges in Complexity and Informatics". *Complexity International*, volume 08. <http://www.complexity.org.au/ci/vol08/green05/green05. pdf>.

Grogan, Paul & Tony Proscio. 2001. *Comeback Cities: A Blueprint for Urban Neighborhood Revival*. Boulder: Westview Press.

Guillermo, E. 2001. "You Know Kathy Nguyen" (November 8-15). *Asian Week*. <http://www.asianweek.com/2001_11_09/opinion_emil.html>.

Guth, Alan. 1997. *The Inflationary Universe: The Quest for a New Theory Of Cosmic Origins*. New York: Helix Books.

Gutmann, A., ed. 1994. *Multiculturalism: Examining the Politics of Recognition.* Princeton, NJ: Princeton University Press.

Hall, Peter & Colin Ward. 1998. *Sociable Cities; the Legacy of Ebeneezer Howard.* New York: John Wiley & Sons.

_____. 1988. *Cities of Tomorrow: An Intellectual History of Urban Planning and Design in the Twentieth Century*. Oxford: Basil Blackwell.

Harris, Charles W. & N. Dines, eds. 1998. *Time-Saver Standards for Landscape Architecture—Design and Construction Data*. London, UK: McGraw-Hill.

Harford, Tim. 2006.*The Undercover Economist: Exposing Why the Rich are Rich, the Poor are Poor, and Why You Can Never Buy a Decent Used Car*. Oxford: Oxford University Press.

Hartig, Terry. 2007. "Three Steps to Understanding Restorative Environments as Health Resources." *Open Space People Space*. Catharine Ward Thompson and Penny Travlou, eds. London: Taylor and Francis.

Hedfors, Per. 2003. "Site Soundscapes: Landscape Architecture in the Light of Sound." Department of Landscape Planning, Swedish University of Agricultural Sciences (Uppsala). <http://diss-epsilon.slu.se/archive/00000325/>.

Hempel. L. C. 1996 *Environmental Governance*. Washington, DC: Island Press.

Henley, Jon. 2005. "Paris Drive to Cut Traffic in Centre by 75%." *The Guardian*. London: Guardian Media Group. <http://www.guardian.co.uk/world/2005/mar/15/france.jonhenley>.

*Hermitary.com* (HC). 2009. "Henry David Thoreau" (Janaury 03). <http://www.hermitary.com/solitude/thoreau.html>.

Holling, C. S. 1995. "What Barriers? What Bridges?" *Barriers and Bridges to the Renewal of Ecosystems and Institutions*. L. Gunderson, C. Holling and S. Light, eds. NY: Columbia University Press.

Huffstutter, P.J. & M. Garvey. 2001. "A Quiet Existence Dissolved by Anthrax" (November 01). *Los Angeles Times*. <http://www.latimes.com/news/nationworld/nation/la-110101nguyen.story>.

Jacobs, J. (1964[1961]) The Death and Life of Great American

Jackson, Kenneth. 1985. *Crabgrass Frontier*. UK: Oxford University Press.

James, Elijah M. 1991. *Micro-Economics: A problem Solving Approach*. Canada: Prentice-Hall.

Kaku, Michio. 1994. *Hyperspace: A Scientific Odyssey Through Parallel Universes, Time Warps, and The Tenth Dimension*. Illustrations by Robert O'Keefe. Oxford: Oxford University Press.

Kamin, Blair. 1997. "Rethinking Public Housing" (summer). *Blueprints Magazine*: 4. <http://www.nbm.org/blueprints/summer97/page4/page4.htm>.

Kangas, Steve. 2008. "Ronald Coase and the Coase Theorem" (December 28). <http://www.huppi.com/kangaroo/L-chicoase.htm>.

Kaplan-Solms, K. & M. Solms. 2000. *Clinical Studies in Neuro-Psychoanalysis: Introduction to a Depth Neuropsychology*. London: Karnac Books.

Kariel, Herbert G. 1980. "Mountaineers and the General Public—A Comparison of Their Evaluation of Sounds in a Recreational Environment." *Leisure Sciences*, vol. 3, no. 2: 155-167.

Karr, James R. 1992. *Ecological Integrity: Protecting Earth's Life Support Systems in Ecosystem Health*. R. Costanza, B. G. Norton, and B. D. Haskell, eds. Washington, DC: Island Press.

Kasulis, T.P. 2009. "Zen as a Social Ethics of Responsiveness." *Journal of Buddhist Ethics*. <http://www.buddhistethics.org/13/zse1-kasulis.pdf>.

Kellermann, Bill. 1989 "Discerning the Angel of Detroit" (October 18). *Sojourners*: 16–21.

Keown, Damien. 1996. *Buddhism: A Very Short Introduction*. Oxford: Oxford University Press.

Knoflacher, Hermann. 2006. "A New Way to Organize Parking: The Key to a Successful Sustainable Transport System for the Future." *Environment and* Urbanization, 18 (2): 387–400.

Kohl, Johanna & Vesa Yli-Pelkonen. 2005. "The Role of Local Ecological Knowledge in Sustainable Urban Planning: Perspectives from Finland" (March 16). *Sustainability: Science, Practice, & Policy*, 1(1):3-14. <http://ejournal.nbii.org/archives/ vol1iss1/0407-007.yli-pelkonen.html>.

______.2004. "Experts of Forest Biodiversity." The Tenth Biennial Conference of the International Association for the Study of Common Property (IASCP). *The Commons in an Age of Global Transition: Challenges, Risks and Opportunities*. Oaxaca, Mexico: Instituto de Investigaciones Sociales. <http://www.iascp2004.org.mx/downloads/paper_339.pdf>.

Kotzen, Benz & Colin English. 1999. *Environmental Noise Barriers—A Guide to Their Acoustic and Visual Design*. London, UK: Spon. London.

Laurance, William. 1999. "Reflections on the Tropical Deforestation Crisis." *Biological Conservation*, 91 109-117.

Lawless, John. 2007. "City on a Hill: Bi-Weekly Column on D.C. News and Politics" (March 22). *Georgetown Voice*. <http://www.georgetownvoice.com/ 2007/03/22/city-on-a-hill-bi-weekly-column-on-d-c-news-and-politics/>.

*LEDA Database* (LEDA). 2008. "LEDA Measure: License Plate Based Traffic Restrictions, Athens, Greece" (December 18). <http://www.leda.ils.nrw.de/ database/measures/meas0205.htm>.

LeDoux, Joseph E. 1996. *The Emotional Brain: The Mysterious Underpinnings of Emotional life*. New York: Simon & Schuster.

Levin, David Michael, ed. 1993. *Modernity and the Hegemony of Vision*. Berkeley, CA: Univ. of California Press.

Levinas, E. 1987. *Time and the Other and Additional Essays*. Pittsburgh, PA: Duquesne University Press.

Lobbia, J.A. 1999. "Bowery Bummer: Downtown Plan Will Make and Break History" (March 27). *The Village Voice*. <http://www.villagevoice.com/1999-03-16/news/bowery-bummer/>.

Louv, Richard. 2006. *Last Child in the Woods*. Chapel Hill: Algonquin.

Lupton, R. & A. Power. 2004. "The Growth and Decline of Cities and Regions." CASE-Brookings Census Brief No.1.

Mahathera, Suvanno. 2009. "The 31 Planes of Existence" (January 12). <http://www.buddhanet.net/pdf_file/allexistence.pdf>.

Mandlebaum, S. J.; L. Mazza, & R. W. Burchell, eds. 1996. *Explorations in Planning Theory*. NJ: Center for Urban Policy Research.

Marcoux, Alain. 2000. "Population and deforestation" (August). *FAO.org*. <http://www.fao.org/sd/WPdirect/WPan0050.htm>.

Matthews, Christopher. 2009. "Deforestation Causes Global Warming." *FAO.org*. <http://www.fao.org/newsroom/en/news/2006/1000385/index.html>.

McAleer, Neil. 1987. *The Mind-Boggling Universe*. New York: Doubleday & Co.

McHarg, Ian. 1975. *Design with Nature*. New York: Wiley, John and Sons.

McNamee, Kevin. 1996. "Accelerating and Enhancing Our Efforts to Protect National Heritage." *Environments*, 24 (1): 35-38.

*Merriam-Webster's Collegiate Dictionary Online* (MWD). 2009. "Mores" (January 11). <Http://www.merriam-webster.com/dictionary/mores>.

_____.2009a. "Norms" (January 11). <http://www.merriam-webster.com/dictionary/norms>.

_____.2008. "Density" (December 15). <http://www.merriam-webster.com/dictionary/density>.

_____.2008a. "Void" (December 15). <http://www.merriam-webster.com/dictionary/void>.

_____.2008b. "Sustainability" (December 15). <http://www.merriam-webster.com/dictionary/sustainability>.

Michaels, David; Stephen R. Zoloth; Phil Alcabes; Charles A. Braslow & Steven Safyer. 1992. "Homelessness and Indicators of Mental Illness Among Inmates in New York City's Correctional System" (February). *Hosp Community Psychiatry*, 43:150-155. <http://psychservices.psychiatryonline.org/cgi/content/abstract/43/2/150>.

Misselwitz, Phillip. 2003. "Military Operations as Urban Planning" (August 27). *Mute: Culture and Politics after the Net.* <http://www.metamute.org/en/military_operations_as_urban_planning>.

Mogato, Manny. 2008. "U.N. Calls on Asian Nations to End Deforestation" (Friday, June 20). <http://www.reuters.com/article/environmentNews/idUSMAN18800220080620>.

Moore Milroy, B. 1991 "Into Postmodern Weightlessness." *Journal of Planning Education and Research*, 10(3): 181-212.

Morris, Eleanor Smith. 1997. *British Town Planning and Urban Design*. England: Longman.

Morris, A.E.J. 1972. *History of Urban Form: Prehistory to the Renaissance*. England: Geroge Godwin.

MSNBC.com. 2005. "Harsh Urban Renewal in New Orleans: Poor, Black Residents Cannot Afford to Return, Worry City Will Exclude Them" (October 12). <http://www.msnbc.msn.com/id/9678167/>.

Mulligan, B; L. Edward, Lewis; M. Flaupel; L. Goodman & L. Anderson. 1987. Enhancement and Masking of Loudness by Environmental Factors—Vegetation and Noise." *Environment and Behavior*, vol. 19, no. 4: 411-443.

Mumford, Lewis. 1967 *The Myth of the Machine: Technics and Human Development*. New York: Harcourt, Brace & Word.

Mweninguwe, Raphael. 2005. "Massive Deforestation Threatens Food Security" (February 15). *News from Africa*. <http://www.newsfromafrica.org/newsfromafrica/articles/art_9607.html>.

Nagao, Gadjin. 1991. *Madhyamika and Yogachara*. Leslie S. Kawamura, tr. Albany, NY: SUNY Press.

Najam, Adil. 2008. The session on “Sustaining Cities: Environment, Economic Development, and Empowerment” (April 04). *M.I.T. World.* <http://mitworld.mit.edu/video/574>.

NASA. 2004. “NASA Data Shows Deforestation Affects Climate In the Amazon” (June 09). <http://www.nasa.gov/centers/goddard/news/topstory/2004/0603 amazondry.html>.

*National Coalition for the Homeless* (NCH). 2008. “Mental Illness and Homelessness” (June). NCH Fact Sheet #5. Washington, DC: National Coalition for the Homeless.

_____. 2004. “Illegal to Be Homeless: 2004 Report.” <http://www.national homeless.org/publications/crimreport2004/introduction.html>.

*National Resource and Training Center on Homelessness and Mental Illness* (NRTC). 2003. “Get the Facts.” <http://www.nrchmi.samhsa.gov>.

*Natural Resources Defense Council* (*NRDC*). 1996, “Breath-Taking: Premature Mortality Due to Particulate Air Pollution in 239 American Cities” (May).

Neumaier-Dargyay, E.K., tr. 1992. *The Sovereign all-creating mind--the motherly Buddha*. Albany: State University of New York Press.

*NHCHC.org* (NHCHC). 2000. “Mental Illness, Chronic Homelessness: An American Disgrace” (October). *Healing Hands*. <http://www.nhchc.org/ healinghands.html>.

Nilsson, Sten. 2009. “Do We Have Enough Forests?” (January 05). *ActionBioscience.org*. <http://www.actionbioscience.org/environment/nilsson .html.

Nelson, J. G. & A. Skibicki. 1997. *Georgian Bay Islands National Park Ecosystem Conservation Plan*. Canada: Heritage Resources Centre, University of Waterloo.

_____.& R. Serafin. 1996. “Environmental and Resource Planning and Decision Making in Canada: A Human Ecological and a Civics Approach.” *Canada in Transition: Results of Environmental and Human Geographical Research*. R. Vogelsang, ed.

*Nova Online* (NO). 2000. “Time Can Vary.” <http://www.pbs.org/ wgbh/nova/time/thinktime.html>.

Ong, Walter J. 1982. *Orality and Literacy—The Technologizing of the Word*. UK: Methuen/ Routledge.

Paldanius, J. 1997. “Experiences of Public Participation in Finland.” *Conflict Management and Public Participation in Land Management*. B. Solberg & S. Miina, eds. EFI Proceedings, No.14: 187-195.

Pallasmaa, Juhani. 1996. *The Eyes of the Skin—Architecture and the Senses.* London, UK: Academy Ed.

Panksepp, J. 1998. *Affective Neuroscience: The Foundations of Human and*

*Animal Emotions*. Oxford: Oxford University Press.

Pekarchik, Karin. 2001. "Alphabet City: The ABCs of Gentrification" (June 11). *Business Week*. <http://www.businessweek.com/magazine/content/01_24/b3736044.htm>.

Pezzey, John 1992 *Sustainability: An Interdisciplinary Guide Environmental Values* 1(1992): 321-362.

Pimm, Stuart L; G. Russell; J. Gittleman; T. Brooks. 1995 "The Future of Biodiversity." *Science*, 269: 347-341.

Pinker, Steven. 2002. *The Blank Slate: The Modern Denial of Human Nature*. New York: Viking.

_____.2002a. "The Blank Slate: The Modern Denial of Human Nature" (October 31). *M.I.T. World*. <http://mitworld.mit.edu/video/23>.

Prajnananda, Bhikkhu. 2009. "The Reform of the Sangha." <http://www.buddhistische-gesellschaft-berlin.de/downloads/brokenbuddhanew.pdf>.

Prentice, I.C., et al. 2009. "The Carbon Cycle and Atmospheric Carbon Dioxide" (January 06). <http://www.grida.no/CLIMATE/IPCC_TAR/wg1/pdf/TAR-03.PDF>.

Raman, Shankar. 1997. Lectures on Literary Theory. Cambridge, MA: MIT (spring).

Reid, Lesley W. & Robert M. Adelman. 2003. "The Double-edged Sword of Gentrification in Atlanta." American Sociological Association. <http://www.asanet.org/footnotes/apr03/indexthree.html>.

Richardson, F.; & J. Christopher. 1993. "Social Theory as Practice: Metatheoretical Frameworks for Social Inquiry." *Journal of Theoretical and Philosophical Psychology*, 13:137-153.

Riley, Denise. 2002. "The Right to Be Lonely." *Differences*, 13(1): 1–13.

Roberts, William H. 1970. "Design of Metropolitan Open Space Based on Natural Process." *Metropolitan Open Space and Natural Process*. Philadelphia: University of Pennsylvania.

Robinson, J. B.; G. Francis; R. Legge & S. Lerner. 1990. Defining a Sustainable Society: Values, Principles and Definitions." *Alternatives*, 17(2): 36-46.

Roessner, Jane. 2000. *A Decent Place to Live: From Columbia Point to Harbor Point*. Boston: Northeastern University Press.

Rosser, J. Barkley, ed. 2003. *Complexity in Economics*. <http://cob.jmu.edu/rosserjb/COMPLEXITY%20IN%20ECONOMICS.doc>.

Sacks, O. 1984. *A Leg to Stand On*. New York: Summit Books/Simon and Schuster.

Sahlins, M., 1976. *The Use and Abuse of Biology: An Anthropological Critique of Sociobiology*. Ann Arbor: The University of Michigan Press.

Sandel, Michael. 1982. *Liberalism and the Limits of Justice*. New York:

Cambridge University Press/

*ScienceDaily* (SD). 2009. "Deforestation." <http://www.sciencedaily.com/articles/d/deforestation.htm>.

______.2009a. "Confirmed: Deforestation Plays Critical Climate Change Role" (May 11). <http://www.sciencedaily.com/releases/2007/05/070511100918.htm>.

______.2009b. "Rainforest Biodiversity Shows Differing Patterns" (August 14). <http://www.sciencedaily.com/releases/2007/08/070808132022.htm>.

______.2008. "Drivers Of Tropical Deforestation Are Changing, Say Scientists" (August 07). <http://www.sciencedaily.com/releases/2008/08/080805192729.htm>.

Shaw, Jane. 2002. "Public Choice Theory." *The Concise Encyclopedia of Economics*. <http://www.econlib.org/library/Enc/PublicChoiceTheory.html>.

Shlain, Leonard. 1991. *Art and Physics: Parallel Visions in Space,Time, and Light*. New York: William Morrow and Co., Inc.

Shoup, Donald C. 2005. *The High Cost of Free Parking*. Washington, DC: American Planning Association.

Simmel, G. 1997. "The Metropolis and Mental Life." *Simmel on Culture*. D. Frisby and M. Featherstone, eds. London: Sage.

______.1997a. "Bridge and Door." *Simmel on Culture*. D. Frisby and M. Featherstone, eds. London: Sage.

Solms, M. & O. Turnbull. 2002. *The Brain and the Inner World: An Introduction to the Neuroscience of Subjective Experience*. New York: Other Press.

Springgate, Lee. 2008. "Defining Parks and Park Systems." *From Recreation to Re-creation*. American Planning Association.

Sterman, John. 2008. "Why Bad Things Happen to Good Technologies" (January 30). *M.I.T. World*. <http://mitworld.mit.edu/video/553>.

Stock, Jocelyn & Andy Rochen. 2009. "The Choice: Doomsday or Arbor Day" (January 05). <http://www.umich.edu/~gs265/society/deforestation.htm>.

Stokols, Daniel. 1978. "A Typology of Crowding Experiences." *Human Response to Crowding*. Andrew Baum & Yakov M. Epstein, eds. Hillsdale, Michigan: Erlbaum.

Susskind, L. & S. McCreary. 1985. "Techniques for Resolving Coastal Resource Management Disputes Through Negotiation." *American Planning Association Journal,* 51: 365-374.

Tambiah, Stanley Jeyaraja. 1990. *Magic, Science, Religion and the Scope of Rationality*. Cambridge: Cambridge University Press.

Taylor, C. 1992. *Multiculturalism and "the Politics of Recognition."* Princeton, NJ: Princeton University Press.

Thabit, Walter & Frances Fox Piven. 2005. *How East New York Became a Ghetto*.

New York: New York University Press.

Thaler, Richard H. & Cass R. Sunstein. 2008. *Nudge: Improving Decisions about Health, Wealth, and Happiness*. New Haven: Yale University Press.

Thanissaro, Bhikkhu, tr. 2004. "Introduction to the Avyakata Samyutta." <http://www.accesstoinsight.org/tipitaka/sn/sn44/sn44.intro.than.html>.

_____.1997. "Aggi-Vacchagotta Sutta To Vacchagotta on Fire." <http://www.accesstoinsight.org/tipitaka/mn/mn.072.than.html>.

Thera, Piyadassi, tr. 1999. "Dhammacakkappavattana Sutta: Setting in Motion the Wheel of Truth." <http://www.accesstoinsight.org/tipitaka/sn/sn56/sn56.011.piya.html>.

*ThinkQuest.org* (TQ). 2009. "Tropical Rain Forest" (January 05). <http://library.thinkquest.org/11353/trforest.htm>.

Thompson, Ian H. 2000. "The Ethics of sustainability." *Landscape and Sustainability*. John F. Benson and Maggie H. Roe, eds. London, UK: Spon.

Thompson, J. Phillip. 2008. The session on "Sustaining Cities: Environment, Economic Development, and Empowerment" (April 04). *M.I.T. World*. <http://mitworld.mit.edu/video/574>.

Thorne, Kip S. 1994. *Black Holes and Time Warps: Einstein's Outrageous Legacy*. NY: W. W. Norton.

Tomas, Benet. 1997. "Assessment of Deforestation by Acid Rain." XI World Forestry Congress (October 12-22). Antalya, Turkey. <http://www.fao.org/forestry/docrep/wfcxi/publi/V1/T5E/3-43.HTM>.

Tonkiss, Fran. 2003. "The Ethics of Indifference: Community and Solitude in the City" (June 3). *International Journal of Cultural Studies*. <http://ics.sagepub.com/cgi/content/abstract/6/3/297>.

Truax, Barry. 1984. *Acoustic Communication*. New Jersey: Ablex.

Turner, Tom. 1998. "Twelve Alternatives." *Landscape Design*, no. 267: 42-45.

*UN-HABITAT* (UNH). 2007. "What are Slums and Why Do They Exist?" (April 16-20). <http://www.unhabitat.org/downloads/docs/4625_51419_GC%2021%20What%20are%20slums.pdf>.

*United States Environmental Protection Agency* (USEPA). 1994. "Health Assessment Document for 2,3,7,8-Tetrachlorodibenzo-P-Dioxin (TCDD) and Related Compounds" (August, Volume III). Washington DC: EPA/600/BP-92/001c. External Review Draft.

Visser, Matt 1995. *Lorentzian Wormholes: From Einstein to Hawking*. Woodbury, N.Y.: American Institute of Physics.

Wachs, M. 1995 *Foreword in Planning Ethics: A Reader in Planning Theory, Practice and Education*. S. Hendler, ed. Center for Urban Policy Research.

Wallace, Deborah & Rodrick Wallace. 2002. *A Plague on Your Houses: How New York Was Burned Down and National Public Health Crumbled*. New York:

Verso.

Wallace, R. 1990. "Urban Desertification, Public Health and Public Order: 'Planned Shrinkage', Violent Death, Substance Abuse and AIDS in the Bronx." *Social Science & Medicine*, vol. 37, no. 7: 801-813.

*Washington Post, The* (WP). 2001. "Evolution Idea: Flatter's Better" (Monday, September 10): A9.

Wasylenki, Don & George Tolomiczenko. 1998. "Pathways Project, Method and Prevalence Findings." *Proceedings and Recommendations on "Mental Illness and Pathways into Homelessness: Findings and Implications"* (January 16). Mental Health Policy Research Group. <http://www.camh.net/ hsrcu/html_documents/pathways_proceedings.html>.

Weber, J.R. & C.S. Word. 2001. "The Communication Process as Evaluative Context: What Do Nonscientists Hear When Scientists Speak?" *BioScience*, 51: 487-495.

Weber, Max. 1930. *The Protestant Ethic and the Spirit of Capitalism*. Talcott Parsons, tr. New York: Scribner.

Weizman, Eyal. 2003. "Military Operations as Urban Planning" (August 27). *Mute*: *Culture and Politics after the Net*. <http://www.metamute.org/en/military_operations_as_urban_planning>.

_____. & Rafi Segal. 2003a. *A Civilian Occupation: The Politics of Israeli Architecture*. New York: Verso.

Wenger, J. 2001. "From the Wild West" (November 20-26). *The Asian Reporter*, 11(47): 7.

Werner, Karel. 1989. "Mysticism and Indian Spirituality." *The Yogi and the Mystic*. Karel Werner, ed. London, UK: Curzon Press.

*Wikipedia* (WK). 2009. "Deforestation" (January 06). <http://en.wikipedia.org/wiki/Deforestation>.

_____.2009a. "Urban Open Space" (January 08). <http://en.wikipedia.org/wiki/Urban_open_space>.

_____.2009b. "Open Space Reserve" (January 08). <http://en.wikipedia.org/wiki/Open_space_reserve>.

_____.2009c. "Green Belt" (January 08). <http://en.wikipedia.org/wiki/Green_belt>.

_____.2009d. "Public Space" (January 08). <http://en.wikipedia.org/wiki/Public_space>.

_____.2009e. "Urban Decay" (January 09). <http://en.wikipedia.org/wiki/Urban_decay>.

_____.2009f. "Slums" (January 09). <http://en.wikipedia.org/ wiki/Slums>.

_____.2009g. "Planned Shrinkage" (January 09). <http://en.wikipedia.org/wiki/Planned_shrinkage>.

_____.2009h. "Redlining" (January 09). <http://en.wikipedia.org/ wiki/Redlining>.

_____.2009i. "Gentrification" (January 09). <http://en.wikipedia.org/ wiki/Gentrification>.

_____.2009j. "Nothing" (January 11). <http://en.wikipedia.org/ wiki/Nothing>.

_____.2009k. "Buddhism" (January 11). <http://en.wikipedia.org/wiki/ Buddhism>.

_____.2009l. "Talk: Criticism of Buddhism" (January 12). <http://en.wikipedia.org/wiki/Talk:Criticism_of_Buddhism>.

_____.2009m. "Emptiness" (January 12). <http://en.wikipedia.org/wiki/ Emptiness>.

_____.2009n. "Satisfying" (January 15). <http://en.wikipedia.org/wiki/ Satisfying>.

_____.2008. "Posthumanism" (October 20). <http://en.wikipedia.org/ wiki/ Posthumanism>.

_____.2008a. "Transhumanism" (October 20). <http://en.wikipedia.org/wiki/ Transhumanism>.

_____.2008b. "Urban Planning" (December 14). <http://en.wikipedia.org/ wiki/Urban_planning#History>.

_____.2008c. "Grid Plan" (December 14). <http://en.wikipedia.org/ wiki/Grid_plan>.

_____.2008d. "Hippodamian of Miletus" (December 14). <http://en.wikipedia.org/ wiki/Hippodamian>.

_____.2008e. "Medieval Fortification" (December 15). <http://en.wikipedia.org/ wiki/Medieval_fortification>.

_____.2008f. "Star Fort" (December 15). <http://en.wikipedia.org/ wiki/Star_fort>.

_____.2008g. "Garden City Movement" (December 15). <http://en.wikipedia.org/ wiki/Garden_city_movement>.

_____.2008h. "Greenbelt" (December 15). <http://en.wikipedia.org/ wiki/Greenbelt>.

_____.2008i. "Brasilia" (December 15). <http://en.wikipedia.org/wiki/ Brasilia#cite_note-1>.

_____.2008j. "Strategic Urban Planning" (December 15). <http://en.wikipedia.org/ wiki/Strategic_Urban_Planning>.

_____.2008k. "Traffic Congestion" (December 18). <http://en.wikipedia.org/ wiki/Traffic_congestion>.

_____.2008l. "Road Rage" (December 26). <http://en.wikipedia.org/ wiki/Road_rage>.

_____.2008m. "Sigmund Freud" (December 26). <http://en.wikipedia.org/

wiki/Sigmund_Freud>.

______.2008n. “Descartes’ Error” (December 26). <http://en.wikipedia.org/wiki/Descartes%27_Error>.

______.2008o. “Principles of Intelligent Urbanism” (December 28). <http://en.wikipedia.org/wiki/Principles_of_Intelligent_Urbanism>.

______.2008p. “Urban Renewal” (December 29). <http://en.wikipedia.org/wiki/Urban_Renewal>.

______.2008q. “Lowest Common Denominator” (March 26). <http://en.wikipedia.org/ wiki/Least_common_denominator>.

______.2008r. “Infinity” (November 11). <http://en.wikipedia.org/wiki/Infinity>.

______.2008s. “Transfinite Number” (November 10). <http://en.wikipedia.org/wiki/Transfinite_ number>.

______.2008t. “Quantum Mechanics” (November 26). <http://en.wikipedia.org/wiki/Quantum_mechanics>.

______.2008u. “Copenhagen Interpretation” (November 26). <http://en.wikipedia.org/wiki/Copenhagen_interpretation >.

______.2008v. “Many Worlds Interpretation” (November 26). <http://en.wikipedia.org/wiki/Many_Worlds_Interpretation>.

______.2008w. “Hyperspace (Science Fiction)” (November 23). <http://en.wikipedia.org/wiki/Hyperspace_(science_fiction)>.

______.2008x. “Time Travel” (November 26). <http://en.wikipedia.org/wiki/Time_travel#Time_dilation>.

______.2008y. “Wormhole” (November 25). <http://en.wikipedia.org/wiki/Wormhole>.

______.2007. “Linguistics” (March 04). <http://en.wikipedia.org/wiki/Linguistics>.

______.2007a. “Government Failure” (March 14). <http://en.wikipedia.org/wiki/Government_failure>.

*Wikiversity* (WKV). 2008. “Why 10 Dimensions” (November 23). <ttp://en.wikiversity.org/wiki/Why_10_dimensions>

Wilczek, Frank. 2008. “The Lightness of Being: Mass, Ether, and the Unification of Forces: Anticipating a New Golden Age” (October 16). *M.I.T. World*. <http://mitworld.mit.edu/video/618>.

Wilke, Claus, et al. 2001. “Evolution of Digital Organisms at High Mutation Rates Leads to Survival of the Flattest.” *Nature* (July 19).

Wirth, L. 1995. “Urbanism as a Way of Life.” *Metropolis: Centre and Symbol of Our Time*. P. Kasinitz, ed. Basingstoke: Macmillan.

Wittgenstein, Ludwig.1953. *Philosophical Investigations*. G. E. M. Anscombe and R. Rhees, eds. G.E.M. Anscombe, tr. Oxford: Blackwell.

Wolfe, J. M. 1989 “Theory, Hypothesis, Explanation and Action: The Example of Urban Planning.” *Remaking Human Geography*. A. Kobayashi and S.

Mackenzie, eds. Winchester, MA: Unwin Hyman.

Young, I.M. 1990. *Justice and the Politics of Difference*. Princeton, NJ: Princeton University Press.

Young, Terence. 1995. "Modern Urban Parks." *Geographical Review*: 535-551.

Zegras, Chris. 2008. The session on "Sustaining Cities: Environment, Economic Development, and Empowerment" (April 04). *M.I.T. World*. <http://mitworld.mit.edu/video/574>.

Zimbardo, Philip. 2007. "The Lucifer Effect: Understanding How Good People Turn Evil" (April 02). <http://mitworld.mit.edu/video/459>.

_____. 2007a. *The Lucifer Effect: How Good People Turn Evil*. London: Rider.

# INDEX

## •A•

## •B•

## •C•

## •D•



## •E•

## •F•

# •H•

## •K•



## •L•



## •M•

**•N•**

## •O•



## •P•

## Q•



## •R•

## •T•

## •U•

**•V•**



**•W•**



**•X•**



**•Y•**

## •Z•